Adaptations of Desert Organisms

Edited by J.L. Cloudsley-Thompson

Springer
Berlin
Heidelberg
New York
Barcelona
Budapest
Hong Kong
London
Milan
Paris
Santa Clara
Singapore
Tokyo

Volumes already published

Ecophysiology of the Camelidae and Desert Ruminants
By R.T. Wilson (1989)

Ecophysiology of Desert Arthropods and Reptiles
By J.L. Cloudsley-Thompson (1991)

Plant Nutrients in Desert Environments
By A. Day and K. Ludeke (1993)

Seed Germination in Desert Plants
By Y. Gutterman (1993)

Behavioural Adaptations of Desert Animals
By G. Costa (1995)

Invertebrates in Hot and Cold Arid Environments
By L. Sømme (1995)

Energetics of Desert Invertebrates
By H. Heatwole (1996)

Ecophysiology of Desert Birds
By G.L. Maclean (1996)

Plants of Desert Dunes
By A. Danin (1996)

Biotic Interactions in Arid Lands
By J.L. Cloudsley-Thompson (1996)

Structure-Function Relations of Warm Desert Plants
By A.C. Gibson (1996)

Physiological Ecology of North American Desert Plants
By S.D. Smith, R.K. Manson, and J.E. Anderson (1997)

Ecophysiology of Small Desert Mammals
By A.A. Degen (1997)

Homeostasis in Desert Reptiles
By S.D. Bradshaw (1997)

Ecophysiology of Amphibians Inhabiting Xeric Environments
By M.R. Warburg (1997)

Avian Desert Predators
By W.E. Cook (1997)

William E. Cook

Avian Desert Predators

With 25 Figures

Springer

Prof. Dr. William E. Cook
Columbia-Greene Community College
4400 Route 23
Hudson, New York 12534
USA

Front cover illustration:
Drawing of a roadrunner (*Geococcyx californianus*), © James Coe.

ISSN 1430-9432
ISBN 3-540-59262-8 Springer-Verlag Berlin Heidelberg New York

Library of Congress Cataloging-in-Publication Data Cook, William E., 1947- Avian desert predators/ William E. Cook. p. cm. – (Adaptations of desert organisms) Includes bibliographical references (p.) and index. ISBN 3-540-59262-8 (hc) 1. Birds of prey–Adaptation. 2. Desert animals–Adaptation. I. Title. II. Series. QL 696.F3C66 1997 598.9`1742–dc21

Printed in Germany

Cover design: Design and Production GmbH, Heidelberg
Camera ready by: Emanuel Rachl
SPIN 10121141 3137/31 - 5 4 3 2 1 0 - Printed on acid-free paper

Contents

Preface

When John Cloudsley-Thompson asked me to contribute a volume to his series on desert adaptations, he asked for a work on the subject of predators and prey, a topic upon which I had previously published. As an ornithologist, however, I requested that I be allowed to confine the work to avian predators. This being accepted, our concern was that the work not conflict with, but, rather, complement Professor Gordon Maclean's volume on the ecophysiology of desert birds. My book does not focus on avian physiology per se, but on the specific behavioral, anatomical, and physiological adaptations made by particular groups of birds. This volume is organized by devoting a chapter to each group, arranged taxonomically. In this arrangement the text provides a systematic reference to desert adaptations of avian predators.

The common and scientific names of birds discussed in this book follow Sibley and Monroe (1990).

Hudson, New York
Spring 1997

W. E. Cook

Preface

CHAPTER 1

Introduction

1.1 The Desert Environment

Deserts are defined as areas where the evaporation of water greatly exceeds the rainfall. As a result, the land is dry for most of the year and the life forms that inhabit these areas typically possess particular adaptations for survival in an arid habitat. This book is about the adaptations made by a particular group of organisms – carnivorous birds – to arid environments.

A desert may be hot or cold. In fact, the arctic and antarctic regions may be considered deserts due to their lack of available water. However, this volume does not treat adaptations to the polar deserts; rather it is confined to discussions of adaptations made by raptors to hot deserts such as the Sahara and the Kalahari, and to cold deserts (deserts with cold winters) such as the Gobi and the North American Great Basin. For a detailed description of the types and characteristics of deserts, the reader is referred to Cloudsley-Thompson's *The Desert*, (1977).

The 11 major deserts of the world, arranged according to size, are (1) the Sahara, (2) the Australian, (3) the Arabian, (4) the Turkestan, (5) the North American, (6) the Patagonian (South America), (7) the Thar (Asia), (8) the Kalahari and the Namib (Africa), (9) the Takla-Makan (Asia), (10) the Iranian, and (11) the Atacama (South America). These and several smaller deserts are shown in Fig. 1. The Old World deserts are older geologically than the deserts of America, so adaptations for arid habitats among Old World desert organisms would be expected to be at a further stage of development (Bartholomew and Cade 1963, Dawson and Bartholomew 1968, Farner and Serventy 1960).

One Old World desert, the Negev in Israel, illustrates how life has evolved to exist in an area where there appears to be no vegetation at all to support a food chain (Jones and Schachak 1994). The land surface consists of just rocks, including limestone outcrops and small stones, however, lichens thrive in spaces between rock particles hidden to the human eye. Snails graze these endolithic lichens contributing to the production of a small bit of soil and undoubtedly nourish some raptors. Shrikes in North American deserts, for example, are known to feed on snails (Bent 1950).

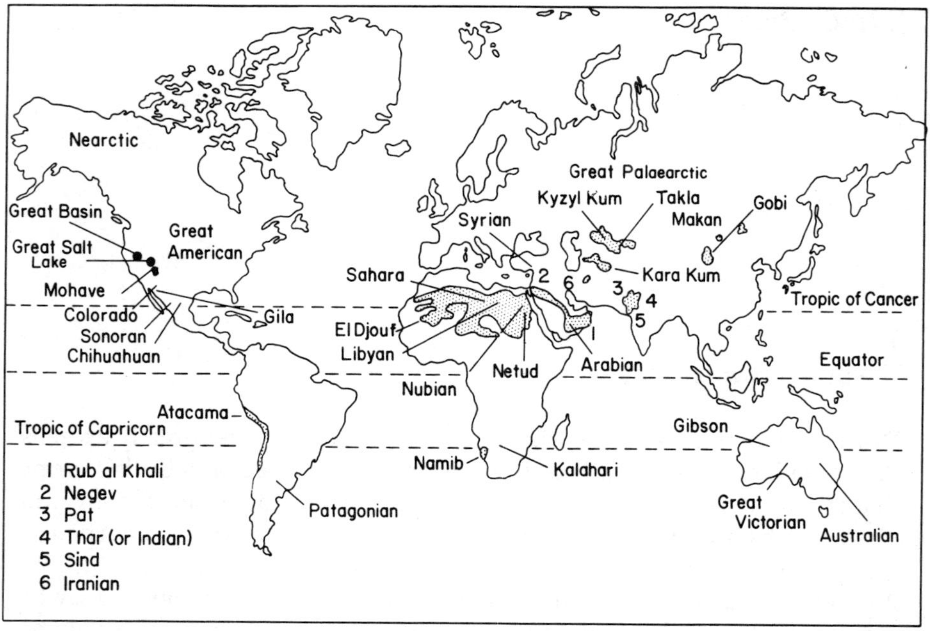

Fig. 1. Location of the world's deserts. (Illustration by John Cloudsley-Thompson)

Curiously, the Australian desert hosts more birds than mammals, a situation that is opposite to that found in other deserts of the world. One might conclude from this that birds have been able to adapt to desert life better than marsupial mammals (Bartholomew and Cade 1963). Elsewhere in the world, however, the volume of desert rodents and lagomorphs (rabbits, hares, and picas) attests to the placental mammals' greater ability to adapt to arid environments.

Some basic elements of avian biology put birds at a disadvantage in deserts. Being typically diurnal, they respire air of low humidity and are exposed to extreme desert heat, and therefore must resort to evaporative cooling and the resulting loss of water. However, other elements of avian biology compensate for these disadvantages. (1) The power of flight allows birds to travel long distances to water holes. (2) A high body temperature facilitates nonevaporative heat loss (Withers and Williams 1990). (3) Birds produce more metabolic water and lose less water in their urine and feces than mammals (Bartholomew and Cade 1963).

Food supplies in desert areas are not always constant. Faced with an inconsistent food supply, birds must be able to switch to eating other foods or move to new locations to find food. Desert birds, therefore, tend to be generalized feeders. Because specialist feeders will tend not to be present, arid areas are characterized by low species diversity. Raptors, on the other hand, tend to be opportunistic in the prey that they select, and find it relatively easy to adapt to deserts (Jaksić and Braker 1983). There will, therefore, be a few regularly occurring species of raptors in the desert, but their population numbers will be highly variable because of the unstable environmental conditions (Poulin et al. 1993). Populations will rise and fall with the food supply.

The unpredictable nature of the food supply has an impact on breeding behavior in desert birds. If there is only enough food to raise a single chick, it is advantageous to allow other hatchlings to die (Soucy 1979; Edwards et al. 1988). Asynchronous hatching assures that chicks vary in size, so that if food supplies are short, the older, larger chick will be fed first and the others will starve, rather than having equal distribution of an insufficient quantity of food to two or more chicks, with the result that all of the nestlings die. Asynchrony may, therefore, be an important adaptation to desert life among raptors.

Ecological constraints also affect postbreeding movements. Juveniles tend to disperse farther in arid areas (Poulin et al. 1993). Fruit and nectar eaters, especially, have to travel to find specialized food. On the other hand, insectivores and raptors may be able to remain more site attached, as their food supply tends to be more consistent.

Raptors and other predators have helped to shape the ecology of deserts in terms of prey population stability, species composition, the evolution of antipredator behavior, and the development of the perception of risk among prey species associated with various habitats (Lima and Valone 1991). Raptors can regulate a prey population and promote between-year stability provided that alternative prey are available in excess and that the predator populations are self-regulated (Erlinge et al. 1984). Thus, predatory birds are valuable in reducing the number of rodents and a lower rodent population reduces the risk of overgrazing sparse desert vegetation.

One reason why there are relatively few birds in deserts is the lack of acceptable escape habitat for prey species. A bird's perception of predation risk, along with its particular escape tactic, influences its selection of habitat (Lima 1993). A given habitat may be avoided as too risky if a bird's escape tactic does not match the physical structure of that habitat. For example, highly social birds that escape into the air, such as Horned Larks (*Eremophila alpestris*) (except during the breeding season when they are not social), may be found in exposed habitats like grasslands and deserts, while less social, more solitary birds, such as Wood Thrushes (*Hylocichla mustelina*), select habitats that are more densely vegetated. Raptors have, therefore, impacted desert ecology by limiting the diversity of birds to those that are adapted to the risks associated with inhabiting an exposed environment.

1.2 Temperature Regulation in Desert Birds

Desert birds must contend with some of the harshest conditions on earth. Ambient temperatures frequently exceed the lethal level for birds of 46-48 °C (115-118 °F) (Brook and Birkhead 1991). On the other hand, the dry desert air does not retain heat very well and nighttime temperatures can fall rapidly. Wintertime nights in cold northern deserts can be especially frigid. Birds, like mammals, are homeothermic, maintaining a relatively constant body temperature, however, some birds allow their body temperature to fall several degrees (hypothermia) to reduce heat loss in a cold environment, or allow it to rise several degrees (hyperthermia) to increase heat loss in a warm environment.

The transfer of heat involves three basic processes: radiation, convection, and conduction. Thermal radiation is the movement of heat energy in the form of a wave (electromagnetic radiation to be more precise) through space (i.e., a vacuum) or a medium. It is the only form of heat (physicists would replace the word "heat" with the word "energy," and restrict the word "heat" to mean the transfer of energy) that can move through space, so it is the only form of heat that reaches the earth from the sun. Thermal convection is the transfer of heat by the circulation of heated gas or liquid. Convective heat loss is increased by about 10% for every 1 m/s (meter per second) increase in wind speed (Brook and Birkhead 1991). Thermal conduction is the transmission of heat energy by means of a medium without the movement of the medium – the medium moves in the case of convection. Conduction also differs from radiation – in radiation the movement of the energy passing through the medium does not occur by means of the medium. The basis of conduction is the transfer of kinetic energy by the collision of particles, such as electrons.

Radiation, convection, and conduction are all important for temperature regulation in desert birds. Desert birds both absorb and emit radiant energy, as do all objects in the universe (all those not at absolute zero). Solar radiation helps birds warm up after a cold desert night, and an overheated bird can radiate excess heat to the environment. Wind, even a slight breeze, is useful to desert birds in the removal of heat from the surface of their bodies by convection. Moving water is even better; a river in the desert may be very attractive to some species for temperature regulation. Conduction allows desert birds to dissipate excess heat to the surrounding air as long as the air is cooler than the surface of the bird's body. The rate of conduction of heat from a bird's body depends upon the temperature differential between surface body heat and the ambient air. Desert birds use these processes of heat transfer to help maintain a body temperature within an acceptable range.

1.2.1 Physiological Cooling

Maintaining a body temperature above cold ambient air is easier than keeping it cool in hot surroundings. The metabolism of homeotherms efficiently converts food energy into heat, but physiological mechanisms for cooling are limited. These cooling mechanisms include raising the body temperature (hyperthermia) so heat can be dissipated to the surrounding air, and evaporative cooling from various internal and external body surfaces.

Evaporative cooling requires water loss and, in some cases, an expenditure of energy, so it is to a bird's advantage to avoid using it until absolutely necessary. One method of forestalling the need to resort to evaporative cooling is to reduce internal heat production. Desert birds have made just such a modification. Birds adapted to hot desert regions have lower metabolic rates – they do not produce as much body heat as other birds (Kendeigh and Blem 1974; Hayes and Gessaman 1980). Birds also have a higher body temperature than mammals, an advantage for hot environments, since a high body temperature facilitates the loss of heat by conduction to the ambient air. Desert birds will respond to an increase in ambient temperature with hyperthermia, and the increase in body temperature will allow nonevaporative heat loss to continue, up to a point. There is a thermal maximum that birds can tolerate, but the limit of hyperthermia for desert birds is very near their lethal temperature – they maximize conductive heat loss before resorting to evaporative cooling.

To further facilitate heat loss by conduction and radiation, the flow of warm blood to surface areas is increased, and patches of bare skin are exposed, when a bird is exposed to high temperatures (Arad et al. 1989). The sides of many birds (the area under the closed wing) are thinly feathered so that heat loss can be increased by raising the wings and exposing the skin to a cooling breeze (Calder 1968a; Schmidt-Nielsen 1978).

Temperature regulation in the brain is important as brain tissues are particularly vulnerable to excess heat. A special network of blood vessels, called the rete mirabile ("wonderful net"), exists in certain mammals and birds to conduct heat away from the brain and prevent brain tissues from overheating (Kilgore et al. 1976; Baker 1993).

As valuable as radiation, convection, and conduction are to the temperature regulation of desert birds, these processes fail to provide sufficient cooling under extreme desert conditions. As the ambient temperature nears a bird's body temperature, the processes of radiation and conduction can no longer remove heat from the bird's body. Evaporative cooling is the only physiological process available for reducing the body temperature of homeotherms under these circumstances. Evaporative cooling is costly, however, in terms of water loss as well as energy expenditure. The process involves the evaporation of water from external and internal body surfaces. External evaporative surfaces include many parts of the body, chief among them the legs (Steen and Steen 1965) and naked skin, including the neck in vultures (Arad et al. 1989). Internal evaporative surfaces include the mouth (buccal cavity) and portions of the respiratory system, especially the air sacs.

The movement of air over internal body surfaces of the mouth and respiratory system is accomplished by panting and a rhythmic contraction of the muscles in the throat area (hyoid apparatus), called gular flutter. Panting moves air through the respiratory system and allows evaporative cooling to take place along the surfaces of the air sacs (Fig. 2). Gular flutter requires little energy (Dawson and Bartholomew 1968) but it is a very effective cooling technique involving the evaporation of moisture in the buccal cavity. The rate of gular flutter does not change with changes in temperature, indicating that the bird uses either the most efficient or the most comfortable tempo (Bartholomew et al. 1968).

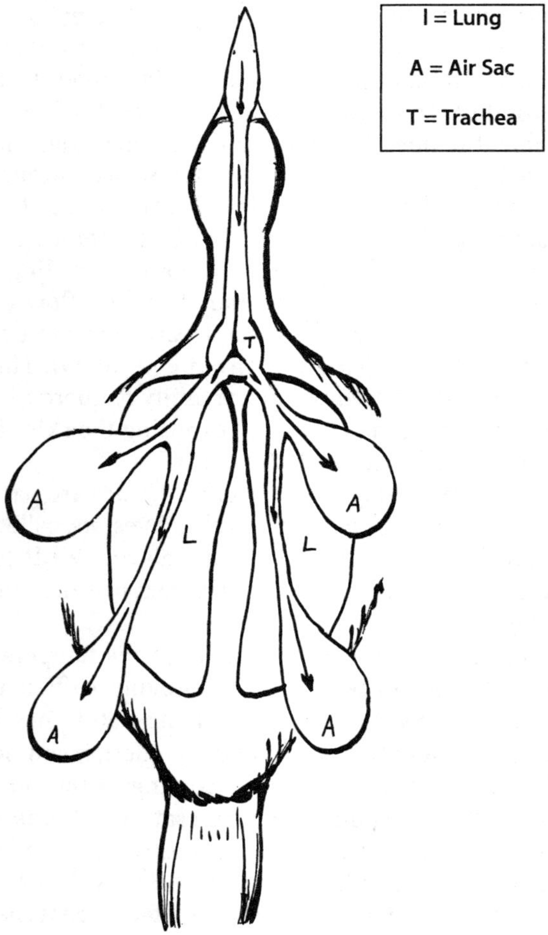

Fig. 2. A bird's respiratory system is connected to large air sacs where evaporative cooling can take place. This air can bypass the lungs so that excessive gas exchange is avoided. L lung; A air sac; T trachea. (Illustration by Will Yandik)

Despite the cooling efficiency of panting and gular flutter, they are only used as a last resort, following evaporation from external body surfaces, because they do require an expenditure of energy (Withers and Williams 1990). Yet, even these physiological cooling processes cannot dissipate enough heat for a bird to tolerate the hottest desert conditions, so relief must be sought in the form of behavioral adjustments to avoid heat gain.

1.2.2 Behavioral Cooling

Augmenting avian physiological processes of cooling are a suite of behavioral activities designed to avoid desert heat. They include techniques for seeking out pockets of cooler air, reducing insulation, and bathing in water if available (Dawson and Bartholomew 1968).

Finding a cool place is often an immediate response to intense heat among birds. Although most desert birds are neither nocturnal nor fossorial, a heat wave will send Australian birds into mammal burrows seeking shelter from the sun and high temperature (Attwood 1982). Strong powers of flight allow most predatory birds to range over broad areas. Many species not specifically adapted to desert areas can range far into a desert in search of prey and return to roost in cooler habitats (Marti 1976). Many that inhabit deserts are able to escape the intense heat of the desert floor by soaring at altitudes where the temperature is considerably cooler.

At higher ambient temperatures a bird may reduce the effective insulation of its plumage by depressing its feathers, pressing them close to the body, and reducing the amount of air trapped within the plumage (Brook and Birkhead 1991).

The cooling properties of water make it very efficient for the removal of body heat. Desert water holes are attractive to some birds, especially larger species with poor powers of flight such as gallinaceous birds. Water holes allow cooling both by conducting heat away from the body and by evaporative cooling. Predatory birds are not normally among the species found using water holes, however. They seem to be able to cope with desert heat in other ways. New World vultures employ an unusual source of water to cool themselves. When they are heated, they, like their close relatives the storks, direct their urine/feces (in birds, urine and feces are combined in the cloaca and eliminated together) onto their legs in order to provide moisture for evaporative cooling.

1.2.3 Pigmentation

The role of pigmentation in thermoregulation is a complicated phenomenon (Hamilton 1973). Black pigment absorbs solar radiation and can therefore be used as a heat collector, yet black plumage on birds may actually be used to prevent solar heat from penetrating through the feathers to the body (Walsberg et al. 1978). Some desert species, like roadrunners, have black skin to absorb solar radiation after a cool night. This dark skin can then be concealed by pale outer feathers during the warmer parts of the day. Many desert species are paler than their moist habitat counterparts; however, the pale coloration may be more important as an adaptation for camouflage than as an adaptation for retarding solar heat gain.

1.3 Advantages of a Carnivorous Diet for Desert Survival

In waterless regions the base of the food chain may be as simple as wind-blown seeds. In this habitat some insects and reptiles flourish, but birds are scarce. Seed-eating birds are, as expected, the most common, followed by insect-eating passerines. At the top of the food chain are a few raptors feeding on the variety of animal life present (Bartholomew and Cade 1963; Fisher et al. 1972). Of all birds, passerines and raptors seem best able to cope with arid conditions (Hogg et al. 1984).

High respiratory water loss in birds would seem to limit their distribution to areas where water is available (Bartholomew and Dawson 1953). Unlike desert rodents, which rarely drink water, lose little water in respiration, are active at night in low temperatures, spend the day in burrows in high humidity, and produce more metabolic water than they lose by respiration, birds of the desert are diurnal (except for owls and goatsuckers), are exposed to extreme desert conditions, respire air at low humidity, lose more water from their respiratory system than they produce metabolically, and are frequently faced with the necessity for extensive evaporative cooling. Yet, a wide variety of carnivorous birds, diurnal as well as nocturnal, are able to survive in waterless deserts. Even though there is a high rate of evaporative water loss in predatory birds, the body fluids of insects and other animal prey, in conjunction with water produced by the metabolism of food, eliminates the need for drinking water in desert raptors (Irwin 1956; Bartholomew and Cade 1963; Cloudsley-Thompson and Chadwick 1964; Dawson and Bartholomew 1968). Most of the passerines found in desert scrub are insectivorous. Examples from the North American desert scrub include Verdin (*Auriparus flaviceps*) and Rock Wren (*Salpinctes obsoletus*). Larger predatory birds, feeding on vertebrates as well as insects, obtain ample water from their diet as well. Bartholomew and Cade (1963) list many species of hawks, owls, and shrikes that have been shown to be able to maintain or gain weight in captivity without drinking. In an earlier study Bartholomew and Cade (1957) demonstrated that American Kestrels (*Falco sparverius*) could maintain body weight in an ambient temperature of 40 °C (104 °F) when fed on beef heart alone, although they panted continuously and, therefore, sacrificed body water for evaporative cooling.

Carnivores have shorter digestive tracts than herbivores because they can digest their food more quickly. As a result they eat frequently (Gill 1990). The frequent intake of succulent food assures that enough water is consumed to maintain water balance. So, in general, eating meat is a major source of water and allows predatory birds to survive in arid regions.

Desert predators tend to be very indiscriminate in their food preferences (Hils 1985). The primary reason for this is that desert wildlife generally offers very few representatives of any one prey species. Whereas raptors in other habitats may specialize solely in fish, rodents, or some other limited cuisine, a desert predator usually needs both the ability and the desire to catch and eat a wide variety of prey species. The characteristic of being extraordinarily unspecialized in the selection of prey items, called "euryphagy," is an important adaptation of raptors to desert survival. The term "omnivorous" implies the consumption of both animal and plant foods, so it is an inappropriate label for most raptors – they generally confine themselves to animal foods, although there are exceptions.

Generalized feeders, including many desert raptors, have longer intestines than specialist feeders, in order to get the optimum amount of energy from the foods that they eat (Barton and Houston 1993). Digestive efficiency is also greater at higher temperatures, so desert birds that permit hyperthermia can obtain more nutrition from their food. Furthermore, desert birds have a lower metabolic rate and digested food moves through their intestines slowly, so there is more time for nutrients to be absorbed into the blood stream. Barton and Houston (1993) also point out that the higher digestive efficiency of generalized feeders allows them to survive on poorer quality foods such as earthworms and carrion.

Carrion may have as much food value as fresh meat, but it is usually rejected by carnivores. It may contain toxins that make it unpalatable, or even poisonous, or the toxins may impede the digestive process of the predator (Karasov 1990; Barton and Houston 1993). Vultures, however, customarily tolerate putrid carrion that contains bacterial toxins. They even appear to use their fecal matter as a protective antiseptic on their legs while they feed on bacteria laden offal (Kear 1968). Other desert raptors will also consume carrion at times, despite the taste – and birds can taste their food (Brower et al. 1968). Even the Loggerhead Shrike (*Lanius ludovicianus*), a passerine carnivore, is able to eat carrion (Bent 1950). Eating carrion is probably an important adaptation for shrikes since the stored carcasses in their larders may get rather putrefied before being consumed.

One disadvantage of a carnivorous diet is the potential to accumulate dangerous levels of toxic chlorinated hydrocarbons from contaminated prey species. These synthetic poisons can interfere with reproductive physiology by causing the thinning of egg shells, and can retard the development of young birds (Busbee 1977; Mader 1977b). The toxic chemicals spread unchecked through the environment. Raptors in deserts far from agricultural areas where the poisons are applied have been shown to have acquired physiologically damaging levels of DDE (dichlorodiphenyldichloroethylene, a product of the partial breakdown of DDT or dichlorodiphenyltrichloroethane), dieldrin, hexachlorobenzene, heptachlor epoxide, and aroclor.

1.4 Water Conservation in Desert Birds

Birds have three sources of water: (1) water available in the environment, called "free water"; (2) water in the tissues of succulent foods; and (3) water created when food is broken down within cells to release energy, called "metabolic water." Animal tissues contain a great deal of water, so most desert raptors meet all of their water needs through the foods they eat and do not require free water (Berger 1961). The power of flight and rapid movement over land is a great advantage – flight allows birds access to remote watering places, but desert raptors use this advantage more to prey upon the birds that gather at waterholes to drink, than to drink themselves.

Since birds tend to be active during the day, they are more exposed to conditions that require evaporative cooling than nocturnal desert inhabitants. Still, desert birds will reduce evaporative cooling loss by curtailing daytime activities during periods of intense heat, or when deprived of water (Dawson and Bartholomew 1968). Even when exposed to harsh daytime conditions, desert birds lose less water by evaporation than do birds from other habitats. According to Williams (1996), birds from arid environments possess physiological adaptations that result in a statistically lower total evaporative water loss than birds from more mesic environments.

1.4.1 Water Conservation in the Respiratory System

Water loss by evaporation is the most serious physiological factor limiting bird distribution in arid regions. The amount of water lost from the respiratory system is in fact greater than the amount of water produced metabolically (Bartholomew and Dawson 1953), even though birds produce more metabolic water than mammals due to their higher metabolic rate (Bartholomew and Cade 1963).

The ratio of evaporative water loss to metabolic water produced is greatest in small birds, so shrikes are more at risk than large raptors in waterless places. The small birds that inhabit deserts take up the gauntlet, in part, with a countercurrent cooling system in the nasal and respiratory passages to recapture some water (Gill 1990). Furthermore, small birds can lose 30-50% of their body water by dehydration without ill effect, and, therefore, they are able to survive until their body fluids can be replaced (Dawson and Bartholomew 1968).

1.4.2 Water Conservation in the Kidney

Birds are preadapted to desert life with the production of a dry urine in the form of uric acid rather than urea (Chew 1961). Uric acid can be excreted as a semisolid whereas urea must be excreted as a liquid. The ureters of birds are capable of peristaltic contractions, and produce a mucus lubricant, so they can move the semisolid uric acid efficiently (Brook and Birkhead 1991). Brook and Birkhead also point out that many desert birds have enhanced their water retention abilities by evolving kidneys with an enlarged medulla for the reabsorption of water from the blood. In addition, during periods of extreme dehydration, the amount of urine produced is reduced by limiting the rate of blood filtration in the kidney, thus retaining water (and metabolic waste products) in the circulatory system (Chew 1961).

Although birds are able to excrete nitrogenous wastes with less water than mammals, the kidney of birds cannot excrete salt as efficiently as mammals. Some desert birds, such as roadrunners, have adjusted by improving their salt-excreting abilities (see Sect. 4.3.3.3).

1.4.3 Water Conservation in the Digestive System

Deserts tend to select for anatomical designs that conserve water in the digestive system. For example, the larvae of certain beetles in the Namib Desert have their malpighian tubules (extensions from the gut into the body cavity of insects, normally used to remove wastes from the body cavity) wrapped around their rectum to withdraw water from the rectum (Nicolson 1990). Among birds, it has been found that desert species are consistently better able to absorb water from the chyme in the lumen of the rectum (Thomas 1982). Furthermore, the ceca, a pair of pouch-like organs at the junction of the small intestine and the rectum, also serve to extract liquid from the intestinal contents (Fenna and Boag 1974). There is even evidence that the small intestine of desert birds is modified to retain chyme longer and reabsorb more water, with additional folds of the mucous membranes lining the lumen (Amanova 1975).

Some desert birds, including roadrunners and some hawks, conserve water from the digestive system by transporting salt from the cloaca into their blood, extracting water from the excretory system into the circulatory system osmotically, then excreting the excess salt from the circulatory system through nasal glands.

1.4.4 Breeding Cycle Adjusted to Rainfall

Water is crucial to successful breeding for birds. Egg production alone requires additional water in the diet for the egg itself (Bartholomew and Cade 1963). Also, the growth of vegetation and production of food, in the form of expanded insect and small mammal populations, following a rainfall is necessary for feeding nestlings. To cope with these constraints, desert birds modify their breeding cycle to fit rainfall patterns. In Australia some species will breed at any time of the year following a sufficient rainfall or, conversely, they will delay breeding for several seasons during a prolonged drought (Keast and Marshall 1954; Farner 1967; Dawson and Bartholomew 1968). North American species are not so flexible but populations have adapted to local areas, breeding in late winter or early spring in deserts with winter rainfall, and breeding in late spring or early summer in deserts with summer rainfall (Dawson and Bartholomew 1968).

African raptors also show evidence of adjusting their breeding season on the basis of rainfall patterns. Some widespread African shrikes exhibit geographical variation in their breeding season, so that different populations can breed during the rainy season in various parts of the continent (Cooper 1970a). Studies of particular African falcons and vultures indicate that some species are able to modify their breeding season in response to rainfall and subsequent food supply (Houston 1990; Kemp 1991).

CHAPTER 2

Vultures

The seven species of New World vultures and the 15 species of Old World vultures represent two separate branches of avian evolution that have converged to form ecologically and morphologically similar birds. The Old World vultures are related to hawks and eagles (Family Accipitridae) while the closest relatives to the New World vultures are storks. Sibley and Monroe (1990) place both the New World vultures and storks in the Family Ciconiidae. It is interesting to note that the proponents of DNA hybridization systematics have lumped into one order (Ciconiiformes) such traditionally diverse groups as shorebirds, gulls, and alcids; hawks, eagles, and vultures; herons, bitterns, egrets, storks, spoonbills, and flamingos; shearwaters, petrels, and albatrosses; pelicans and cormorants; loons; and penguins.

Vultures are not uncommon in deserts. In fact, several of the Old World vultures are characteristic of desert areas. The greatest diversity can be found in Africa on the southern edge of the Sahara. In the deserts and plains of southern Mali, Niger, and Chad one can encounter Egyptian Vulture (*Neophron percnopterus*), Rueppell's Griffon (*Gyps rueppellii*), Lappet-faced Vulture (*Torgos tracheliotus*), and White-headed Vulture (*Trigonoceps occipitalis*). In southern Asia the best-known desert vulture is perhaps the Red-headed Vulture (*Sarcogyps calvus*). In the New World the wide-ranging Black (*Coragyps atratus*) and Turkey Vultures (*Cathartes aura*) can be found in many habitats including deserts.

2.1 Predatory Behavior in Vultures

2.1.1 The Desert Scavenger Niche

In tropical and warm-temperate terrestrial habitats, the ecological role of avian scavenger is typically filled by a vulture. They are usually replaced by other scavengers, such as gulls or jaegers, in cooler latitudes and aquatic regions. Yet despite their specialization for consuming carrion, vultures have customarily been considered raptors. Many vultures, such as the Black Vulture, Andean Condor (*Vultur gryphus*), Egyptian Vulture and White-headed Vulture, do indeed maintain their affiliation to the falconiform clan by occasionally dining on live prey, usually in the form of individuals that are young, weak or otherwise unable to offer much resistance (Stager 1964; Mueller 1967; Karmali 1980; Storer 1988). Some observers have even suggested that Turkey Vultures prefer fresh meat to carrion (Bent 1937; Stager 1964).

Within the scavenger niche, there can be several distinct adaptive strategies. Despite seeming confusion and competition for the same carcasses in East Africa, specialized adaptations allow vultures to partition their resource (Petrides 1959). According to Pennycuick (1972) there are three guilds among the vultures of the Serengeti Plains, each with its own hunting and feeding adaptations.

The first guild to feed includes the Lappet-faced Vulture (Fig. 3) and the White-headed Vulture. These are territorial birds that hunt by looking for food in their local areas. They arrive at a kill quickly and specialize in tearing up tough parts of a carcass. They also kill young and helpless individuals. In order to hunt in a restricted area these birds have adapted with a large, light body that can stay airborne in weak lift. Lappet-faced Vultures are only found in the driest parts of Africa – in these areas they dominate new-found carcasses. At more than 13 kg (30 lb) they intimidate even the aggressive White-headed Vultures. Lappet-faced and White-headed Vultures are both characterized by an enormous beak that is capable of tearing open the hides of the largest game animals including hippopotamus, elephant, and rhinoceros (Grossman and Hamlet 1964). These birds also use their feet to assist their feeding. Lappet-faced and White-headed Vultures hold down a carcass with their feet while twisting and tearing at the skin and meat with their broad, powerful beaks (Kruuk 1967).

The second guild to feed includes the two griffon vultures, the White-backed Vulture (*Gyps africanus*) and Rueppell's Griffon. These birds ride thermals and follow migratory mammal herds long distances away from nesting areas. They do not hold a hunting territory because the large ungulate carcasses for which they are specialized are transient, infrequent and irregular in distribution, and each carcass feeds a large number of birds (Houston 1974). In areas where there are few large mammals, the griffons fly at a high altitude so they can survey far and wide. They must look for carcasses directly in these areas, but where the ungulate density is high, with herds of wildebeest (*Connochaetes taurinus*), zebra (*Equus burchelli*), and Thomson's gazelle (*Gazelle thomsonii*) literally blanketing the landscape, the griffons fly low and rely on watching other vultures to find food. When they see other vultures descending, the griffons will follow quickly in order to feed before the number of competitors becomes too great. Tucker (1991) has demon-

Fig. 3. The heavy beak of the Lappet-faced Vulture (*Torgos tracheliotus*) is capable of tearing open the hide of an elephant while the Egyptian Vulture (*Neophron percnopterus*), with its diminutive beak, feeds on scraps left by other vulture species. (Illustration © James Coe)

strated that White-backed Vultures can obtain information about wind speed and direction, and descend in such a way as to conserve energy and abort a landing attempt with little or no flapping.

The griffon vultures are specialized to eat soft internal organs, so they benefit from the vultures that have come before them and have opened a carcass. The griffons have long bare necks to reach into the soft internal organs and they have a gutter-shaped tongue with "teeth" pointing backward to help hold and pull out the viscera (Kruuk 1967). Since they must travel widely for food, the two species of griffon vultures possess a higher wing loading (weight per unit wing area) – an important adaptation that allows them speedy cross-country travel.

Hooded Vultures (*Necrosyrtes monachus*) (Fig. 4) and Egyptian Vultures (Fig. 3) make up the third feeding guild. They are smaller-bodied, slower fliers, better able to climb thermals, but poor cross-country travelers. With slender, weak bills they eat scraps left by the other guilds. Egyptian Vultures are well known as being the least discriminating of scavengers (Grossman and Hamlet 1964). Lacking the strong beak of a Lappet-faced Vulture, they are forced to feed on carrion scraps and all sorts of soft animal and plant matter such as dates, termites, small crustaceans in water, rats, other small mammals, and human excrement. They can also be found around human habitation picking through garbage, and they patrol burned grasslands looking for scorched animals.

As often is the case, a desert-adapted bird must be capable of eating a wide variety of foods because the desert habitat tends not to offer an abundance of any one food source. Egyptian Vultures are quick to adapt to novel food sources. In Spain, where breeding densities are high enough for nestlings to watch adjacent nests, young vultures will sometimes join their neighbors for dinner and get fed by adults that are not their parents (Donazar and Ceballos 1990). Some populations of Egyptian Vulture have even learned to use tools to extend their food supply. In East Africa, Jane and Hugo van Lawick-Goodall have observed Egyptian Vultures employ rocks to feed on abandoned Ostrich eggs (Lawick-Goodall 1968). The bill of Egyptian Vultures is much too small and weak to crack open an Ostrich egg, but the vultures have learned to pick up rocks in their bill and throw them down on the eggs. After several blows an egg will break open and the vultures can consume the contents.

In North America the Turkey Vulture may be better adapted to desert foraging than the Black Vulture (Hiraldo et al. 1991). Where both species occur, Turkey Vultures tend to hunt singly while Black Vultures tend to hunt in large groups. Groups of Black Vultures are able to drive Turkey Vultures away from large carcasses but the Black Vulture groups tend to stay in nondesert areas where food is more plentiful and can support many scavengers. In such situations it has been observed that Turkey Vultures eat small prey ignored by the Black Vulture groups. In desert areas, however, where food is scarce, the Black Vulture gangs cannot find sufficient food, so they do not venture there, and

Fig. 4. These Hooded Vultures (*Necrosyrtes monachus*) in Nigeria have slender bills, and along with Egyptian Vultures form a feeding guild adapted to feeding on soft animal and plant materials rejected or overlooked by other feeding guilds. (Photograph by John Cloudsley- Thompson)

the solitary Turkey Vulture can harvest small items and the occasional large carcass. Furthermore, the Turkey Vulture is aerodynamically more efficient than the Black Vulture, so it has the ability to range farther over barren desert areas given an equal amount of expended energy. These adaptations seem to make the Turkey Vulture better adapted to desert survival than the Black Vulture and, indeed, Turkey Vultures are even known to nest in the arid deserts of the southwestern United States (Hatch 1970).

Where no competition from Black Vultures is present, Turkey Vultures appear to modify their hunting behavior according to the habitat (Estrella 1994). The Mapimi desert in Durango, Mexico offers carcasses that are smaller and less easily detected (by smell) than do the nearby forests of La Michilia. Vultures searching the forest are more apt to hunt in groups of three or more birds, and fly at higher altitudes, perhaps as high as 100 m (325 ft) or more, to search for large carcasses. In the desert, however, it is more profitable for individual Turkey Vultures to hunt for small carcasses at low altitudes.

Desert birds, and vultures in particular, are adapted to an unpredictable food supply. They have developed a high digestive efficiency with a long gut and a slow flow rate to get the maximum amount of nutrition from the food that they eat (Houston 1988; Barton and Houston 1993). The digestive system of vultures is well suited to eating a variety of foods. The proventriculus, the anterior part of the stomach – a structure found in birds but not reptiles – is particularly well developed in raptors (Gill 1990). It secretes acidic gastric juices (pH 0.2-1.2) and peptic enzymes that can dissolve bones rapidly. The Lammergeier, or Bearded Vulture (*Gypaetus barbatus*), can dissolve a cow's vertebrae in 2 days.

Vultures, unlike most other raptors, subsist on carrion. Carrion has as much food value as fresh meat but it is rejected by most raptors. Microbes in rotten meat produce toxins that either make the meat unpalatable or impede vertebrate digestive enzymes. It is unclear how vultures are able to tolerate the foul meat that they consume. There may be something special about their immune system (Chaffie 1994) or perhaps they have a protective antiseptic in the gut (Janzen 1977).

2.1.2 Do Vultures Hunt by Smell or Sight?

There has been a lengthy and fascinating debate over the issue of whether vultures hunt by smell or by sight. Stager (1964) provides the most thorough review of the literature on this topic. Stager's own work, which encompasses over 25 years of study, clearly indicates that Turkey Vultures, at least, use olfactory cues to help find food. Apparently several New World vultures have a reasonably well developed sense of smell. Turkey Vultures seem to have the superior ability in this arena, followed by King Vultures (*Sarcoramphus papa*).

Black Vultures seem to have a more modest sense of smell. Cobb (1960) demonstrated that the size of the olfactory bulb of Black Vultures is about in the middle of the range for birds in general. The range for the size ratio of the olfactory bulb to the cerebral hemisphere, for the 63 species that Cobb measured, is from 6 (Black-capped Chickadee *Parus atricapillus*) to 33 (Brown Kiwi *Apteryx australis*). The Black Vulture, with a ratio of 17, falls in the middle of the range along with American Woodcock (*Scolopax minor*) and Adelie Penguin (*Pygoscelis adeliae*). Woodcocks and penguins, like many of their ciconiiform relatives, probably use olfactory cues (albatrosses and petrels are well known for their ability to detect oily animal fats spread on open water) to help find food in mud and water, respectively. Black Vultures might, therefore, be assumed to have some olfactory ability as well. Curiously, observations of Black Vultures indicate that they do not use their sense of smell for locating food, but rather follow Turkey Vultures to find carcasses (Houston 1986). Perhaps their sense of smell simply is not good enough to operate in air several feet above the ground.

In a later study of olfactory bulb size (Bang and Cobb 1968), the Turkey Vulture exhibited the eighth largest olfactory bulb of 108 avian species studied and the largest of all the New World vultures. The King Vulture, of

Central and South America, had the second largest olfactory system of New World vultures. King and Turkey Vultures often hunt in densely forested areas where carrion is hidden from sight, so a good sense of smell would be most beneficial to them (Houston 1986; Proctor and Lynch 1993). Turkey Vultures also hunt open areas by flying at low elevations where odors can be detected.

In contrast, the Old World vultures, hunting the open grasslands and desert borders of Africa, seem to hunt by sight alone (Fig. 5). The eyes of vultures, hawks, and eagles have a special adaptation that serves to increase their visual acuity (Proctor and Lynch 1993). The fovea centralis is a shallow pit with sloping sides in these species. Since the fovea pit has a slightly greater surface area than other areas of the retina, the image is projected over a greater number of photoreceptors. A hunting vulture therefore can see detail at great distances. Lappet-faced Vultures and White-headed Vultures are believed to search for carrion over the African plains by sight alone. White-backed Vultures and Rueppell's Griffons also use visual cues to find food, either directly or by watching for Lappet-faced Vultures and White-headed Vultures that are feeding on a carcass or descending to do so.

Fig. 5. Old World vultures like this Eurasian Griffon (*Gyps fulvus*) hunt by sight alone. (Photograph by John Cloudsley-Thompson)

Some doubt has recently been cast on the Turkey Vulture's ability to smell even the foulest carrion at the height from which they search for food. Smith and Paselk (1986) obtained experimental results indicating that the chemical odorants emanating from a carcass are dispersed rapidly and could not be smelled for long distances. They determined the approximate olfactory response thresholds of Turkey Vultures to three odorants (butanoic acid, ethanethiol, and trimethylamine) associated with animal decomposition. Turkey Vultures could detect butanoic acid and ethanethiol in concentrations as low as 1×10^{-6} (0.000001) moles of odorant per liter of air, but at a typical hunting position of 61 m (200 ft) altitude and 183 m (600 ft) downwind, the concentration of odorant from a typical source was ascertained to be only about 1×10^{-13} (0.0000000000001) moles of odorant per liter of air. Furthermore, the diet of Turkey Vultures, mostly small bodied animals and freshly killed carcasses, does not yield much odorant. Yet, even these authors hesitate to claim that Turkey Vultures cannot hunt by smell since the weight of evidence still strongly suggests that they can.

It seems clear from Stager's studies that Turkey Vultures employ both visual and olfactory cues in their hunting strategy. Turkey Vultures will immediately turn in the direction of a source of released odorant but they will not land on the ground unless they verify the existence of a food source visually. Their olfactory ability seems to give Turkey Vultures an auxiliary sense, useful in detecting carrion, at least at close range. According to Houston (1986), Turkey Vultures will be able to smell a carcass after it has decayed enough to emit an odor – this takes about 1 day. However, they will reject a rotten 4-day old carcass, possibly because of the increased level of bacterial toxins.

There is no dense canopy to conceal carrion in the desert and vultures of deserts, plains, and grasslands apparently do not need to be able to smell their prey. The deserts of the Old World are older than those of the New World (Bartholomew and Cade 1963) so African animals have had more time to evolve desert adaptations. If the ancestors of the African vultures possessed any olfactory ability one might expect that ability to have gone the way of troglodyte eyes. However, it is more likely that the Old World vultures, having probably evolved from a hawk-like ancestor, never had much in the way of a sense of smell. On the other hand, the New World vultures, tracing their lineage to a common ancestor with the storks, may have come equipped with a sense of smell – many shorebirds have olfactory abilities. This olfactory ability has allowed New World vultures to occupy a wider diversity of habitats. Turkey Vultures may be found flying over open grasslands and deserts low enough to smell a carcass, while in neotropical forested areas, Turkey and King Vultures can be found sniffing out rotten tidbits concealed from sight by the vegetation.

2.1.3
Nasal Epithelium

Stager's (1964) research on the olfactory anatomy of vultures revealed an interesting desert adaptation in Turkey Vultures. Compared to King Vultures, which live in a moist forest environment, Turkey Vultures have a large number of mucous glands that can keep their nasal passages moist in arid desert air. Stager found that the cell layer below the olfactory epithelium, a tissue that he believed was the tunica propria, contained a large concentration of mucous glands and speculated that this enhanced ability to keep the olfactory epithelium moist was another feature that enabled Turkey Vultures to inhabit desert areas.

2.1.4
Defense

In certain social circles, human males defend their alcoholic beverages from poachers, or at least claim to, by declaring that they have spit in their beer. This mildly repulsive suggestion is usually offered more for its contribution to the male bonding ritual than for any real resource protection. Vultures, on the other hand, especially in nutrient poor areas like deserts, take resource defense a bit more seriously, and are frequently apt to regurgitate their stomach contents on an assailant (Wagner 1980). Putrefied to begin with, added rarefaction of the carrion in the bird's stomach makes the stomach contents an effective defense against predators as well as a deterrent to many, particularly mammalian, competitors for a food source (many mammals rely on smell to find food).

2.1.5
Breeding Timed to Food Availability

Natural selection tends to refine the breeding cycle of birds so that the young are hatched at the time of maximum food availability. In north temperate songbirds, the majority of young are produced between the end of May and the beginning of July when insects are most abundant. North temperate raptors tend to hatch their young a bit earlier to take advantage of a new crop of mice and voles emerging in March or April. For East African vultures, the time to hatch chicks does not correspond to new life among their prey but rather to the time of greatest hardship and death. At the end of the dry season from August to October, large numbers of migratory wildebeests, zebras and gazelles are dying from malnutrition, thirst and disease. The two most numerous vulture species in the area, the griffon vultures (the White-backed Vulture is the most abundant), time their egg laying so that their fledglings reach their hungriest growth period between August and October (Houston

1976). To this end, the White-backed Vultures lay their eggs in April while the larger Rueppell's Griffons, with a longer incubation period, lay their eggs in January. Houston (1990) has also shown that Rueppell's Griffons can breed at other times of the year when food supplies are available.

The large body of nestling vultures is an advantage in surviving the unpredictable food supply of the desert. Cape Griffons (*Gyps coprotheres*), also known as Cape Vultures, breed in South Africa, including portions of the Namib and Kalahari Deserts (Boshoff and Vernon 1980; Brown et al. 1982). Nestling Cape Griffons were examined and found to have enough body fat to sustain them for possibly as long as 8 days (Komen 1991). Nestlings with large bodies use up energy reserves at a relatively slower rate than nestlings with small bodies. Komen also speculated that nestlings may reduce their metabolic rate during periods of fasting so that energy supplies could last even longer.

2.2 Temperature Regulation in Vultures

The desert can make a warm-blooded person, or raptor, both hot and hungry in short order. When overheated and starving, which need would you attempt to satisfy first? Despite our familiarity with our daily demands for food, some researchers believe that temperature regulation requires more immediate attention than nutrition. Turkey Vultures, deprived of food and water for 36 h and heat-stressed, ignored food and water when returned to their aviary, assumed a spread wing posture, and extended their necks for 10-15 min (Arad and Bernstein 1988). Temperature regulation is perhaps the most important constraint for survival among homeotherms.

Among vultures, Turkey Vultures seem to be particularly able to tolerate a wide range of temperatures without resorting to energy-demanding, metabolic temperature regulation mechanisms. The relatively wide thermoneutral zone of Turkey Vultures allows them to live in diverse habitats including deserts. Further, birds adapted to hot climates also seem to have a lower metabolic rate, so that they produce less body heat and require less energy for cooling.

Turkey Vulture chicks have adapted in such a way as to be able to regulate their body temperature metabolically very early in life (Bent 1937). After the second week of life, long before the appearance of feathers, and while the general development of the body is still quite immature, Turkey Vulture chicks are able to maintain a constant body temperature of 39.8 °C (103.6 °F). Some authors put the thermal independence of Turkey Vulture chicks as early as the age of 5 days (Palmer 1988a). This may be an important adaptation that allows both parents to engage in the time-consuming task of searching for carrion, and not be tied to the nest providing thermal protection for the chick.

2.2.1 Soaring as a Means of Temperature Regulation

On November 29, 1973 a commercial aircraft collided with a Rueppell's Griffon while cruising at an altitude of 11280 m (37000 ft) over Abidjan, Ivory Coast, western Africa (Laybourne 1974). What in the world was the vulture doing up there? Can a vulture see well enough to hunt at that elevation? Could it even see a herd of wildebeest 11.3 km (7 miles) up? Chances are it could detect, at least, the dust cloud of a large mammal herd moving on the plain below. Rueppell's Griffons are known to ride thermals and travel long distances following mammal herds (Pennycuick 1972). But what advantage is there to being so high up? If their vision is good enough, they could be surveying large areas for carcasses, but temperature regulation might be an important consideration as well.

A vulture drifting effortlessly, or seemingly so, over the burning desert sands in search of carrion may not be experiencing the brutality of the desert heat at all. While reptiles and mammals on the desert floor are in near proximity to the harshest of thermal conditions, the air temperature at a vulture's soaring altitude may be 10-15 °C (18-27 °F) cooler (Bartholomew and Cade 1963). Air temperature in the troposphere, which is the lowest layer of the atmosphere and is about 16 km (10 miles) thick at the equator, decreases about 3 °F for every 30 m (1000 ft) in elevation (Beiser 1971). This means that if the desert floor is sizzling at 46 °C (115 °F) the temperature at 2750 m (9,000 ft) would be about 31 °C (88 °F), well within the vulture's zone of thermoneutrality. So at this elevation a vulture could hunt over the hottest desert in relative comfort. But can this explain why a vulture would be at an elevation of 11280 m (37000 ft)? Assuming that the aforementioned altitude and temperature relationship remains consistent, the temperature at 11000 m should be about -16 °C (4 °F). Soaring in such frigid air might be understandable if the bird had spent some time on the desert floor and had a considerable heat load to dissipate and preferred to lose the heat by convection and radiation rather than the evaporation of precious water supplies.

2.2.2 Heat Dissipation and Associated Behavior

In order to avoid overheating, vultures dissipate heat from several regions of the body. The naked skin of the head, neck and legs of Turkey Vultures has large veins, very near the surface, that allow heat to be radiated to the air (Arad and Bernstein 1988; Arad et al. 1989). There are also heat exchangers in the head, wings and legs that can be adjusted to radiate or conserve heat. In these exchangers, heat can be transferred from arteries to nearby veins. This process is particularly important in cooling arterial blood that is flowing to the brain.

As ambient temperatures rise, Turkey Vultures modify their behavior to take advantage of their skin vascularization and heat exchangers (Arad et al. 1989):

- At 30 °C (86 °F) they hold their head horizontally, expose more skin of the head by smoothing out some of the wrinkles, erect their neck feathers to reduce insulation, and move more blood through the veins near the surface of the skin of the head and feet.

- At 40 °C (104 °F) they extend their wings to allow air to circulate between their wings and body, extend their head to expose the bare skin of the neck, and further dilate the vessels of the head and neck to allow more blood to flow through the surface veins of the skin. Black Vultures also expose the bare skin of their neck to increase heat dissipation (Larochelle et al. 1982).

- At 45 °C (113 °F) Turkey Vultures droop their wings, pant with their mouth open and urinate on their legs.

- At 50 °C (122 °F) they pant, move more air over their moist throat (gular flutter) for evaporative cooling, and drink water.

- At 53 °C (127 °F) they hold their wings farther from the body. The ventral side of the wing is believed to be an important site of heat dissipation but the skin is not as highly vascularized as in the head and legs. Still, during flight and soaring, convection of heat from the large superficially located veins of the wings would enhance body cooling.

2.2.3 Legs as Sites of Temperature Regulation

In 1967 David Ligon suggested that the New World vultures were related to storks based on skeletal and behavioral similarities. Among them was the trick of urinating on the legs to cool the body by evaporation. Kahl (1963) had described this behavior in Wood Storks (*Mycteria americana*) and Black Vultures. "Hidrosis" is the technical term for sweating so Kahl called this behavior "urohidrosis." Ligon (1967) and Hatch (1970) later reported urohidrosis in Turkey Vultures as well. Furthermore, Hatch indicated that urohidrosis is a very effective cooling technique and was used in conjunction with panting to lower the body temperature of Turkey Vultures. The salient point of Hatch's work was that the evaporation of a small amount of water, from such a small area of a bird's body as represented by the legs, can have a significant effect upon the temperature of the whole body.

Urohidrosis is such a splendid example of taking advantage of and recycling available resources that it is almost surprising that other desert birds have not adopted this mechanism. The genetic basis of urohidrosis apparently appeared recently, in only one branch of avian evolution, so only the descendants of that line have either the ability or the inclination to respond to thermal stress by urinating on the legs, until the unlikely event that the gene, or combination of genes, appears in another group of birds. In the course of evolution, it would be more likely that a descendant of a vulture, equipped for desert survival with urohidrosis, would be modified to fill a desert niche, than that another desert bird would spontaneously get the mutant gene or gene combination necessary to perform urohidrosis.

For birds in general, the legs are important sites of thermoregulation, with or without urohidrosis. Blood flow through a network of vessels in the legs can be regulated to allow for the conservation of heat during cold weather and for heat dissipation during hot weather (Steen and Steen 1965). Gulls, for example, can stand on ice and not lose much heat, because the arterial blood going to the feet is shunted through a series of vessels lying next to veins returning cool blood. The blood going to the feet is cooled and the blood returning to the body is warmed. However, this countercurrent heat conservation system is bypassed in hot desert situations. Here, warm blood flowing to the feet is directed to vessels nearer the surface of the skin where it can dissipate some of its heat load to the environment. Furthermore, in high ambient temperatures, blood flow is speeded up so that less heat exchange takes place between arterial and venous blood and therefore less heat is returned to the body.

2.2.4 Hyperthermia

Black Vultures allow their body temperature to rise when the ambient temperature rises (Larochelle et al. 1982). In this way they do not have to expend as much energy for temperature regulation. As long as the body temperature is above the ambient temperature, heat will flow away from the bird's body. Even when the ambient temperature is higher than body temperature, allowing the body temperature to rise will result in less expenditure of energy for cooling. Temperature regulation is very costly, so Black Vultures will avoid it, especially when food supplies are scarce and they are forced to endure a period of fasting.

2.2.5 Physiological Adaptations for Heat Conservation

The desert can be surprisingly cold at times. Even in the hot Sonoran Desert, in the summer, night time temperatures drop to 12.5 °C (54.5 °F) (Clark and Ohmart 1985). The Great Basin Desert of Colorado can be -30 °C (-22 °F) in mid March when Turkey Vultures are beginning to arrive from their southern winter range (Millsap and Millsap 1987). Adaptations to desert cold include hypothermia, good insulation, moderate increase in metabolic rate, countercurrent heat exchangers to conserve heat dissipation from the circulatory system, and behavioral modifications.

Contrary to the belief that hypothermia only occurs in small animals, Turkey Vultures allow their body temperature to drop 6 °C (11 °F) from 40 °C (104 °F) during the day to 34 °C (93 °F) during the night (Gill 1990). In hibernating rodents hypothermia is a technique to conserve energy during cold winters when food availability is reduced. Vultures have adapted hypothermia to conserve energy during cold nights (Heath 1962; Hatch 1970). This adaptation may be especially important to vultures in deserts, where food is not plentiful and where nights can be very cold.

Turkey Vultures have been shown to be well insulated by exhibiting a low heat transfer during cold exposure (Arad and Bernstein 1988). A moderate increase in metabolic rate produces additional body heat and helps this vulture species adapt to cool as well as warm habitats, and to survive cold desert nights.

The Turkey Vulture has countercurrent heat exchangers in the legs, wings, and head that reduce heat loss in cold environments (Arad et al. 1989). Warm arterial blood, flowing toward the surface, can be shunted through a network of blood vessels, called a plexus or rete, where many arteries lie in close association with a network of veins bringing cool blood back from the extremities. In the rete, heat is transferred from the arteries to the veins and is thereby conserved as it is carried back to the interior of the body.

2.2.6 Behavioral Modifications for Heat Conservation

Most North American Turkey Vultures are migratory. They move south to warmer regions in the winter, yet they possess a remarkable ability to cope with cold weather. They seek shelter in caves, hold their wings close to their body to increase insulation, retract their naked necks into their feathered collars, wrinkle the skin of their head to reduce exposure of surface blood vessels, and reduce blood flow to the skin of the head and feet (Arad et al. 1989). Turkey Vultures also spread their wings to make use of the warming rays of the morning sun (Clark and Ohmart 1985). Sunbathing is useful for both warming after a cool night and for drying the feathers after rain. In the Sonoran Desert it rains on about 15% of the days of the year so the potential for nocturnal hypothermia and the consequent need for drying feathers even in a desert environment are evident.

2.3 Water Conservation in Vultures

2.3.1 Sources of Water

Birds obtain water from three sources: drinking free water, eating food with high water content, and the production of water through cellular respiration, called metabolic water.

The capacity to move rapidly over long distances, then return to water, has been considered an important adaptation associated with desert homeotherms (Petrov 1975). One might assume that vulture species that are most capable of long-distance soaring, such as White-backed Vultures and Rueppell's Griffons in Africa south of the Sahara, would be the most frequently found vultures in desert areas. This is not always the case, however. Rueppell's Griffons are found in the desert but White-backed Vultures tend to remain in the less arid plains. On the other hand, many desert vultures that are local territory holders and do not travel long distances for any reason, not even to obtain water, are found in the desert. Lappet-faced Vultures and White-headed Vultures are southern Sahara examples. These birds apparently obtain all the water they need from their diet. This is quite understandable because they tend to be the first scavengers to arrive at a carcass. They are able to open the carcass and feed on fresh carrion that has had little time to become desiccated and is therefore still quite succulent.

Larger predatory birds are generally able to obtain adequate quantities of water from their food (Bartholomew and Cade 1963). They will drink water when it is available but they do not need it to survive. Daniel Hatch (1970) kept a Turkey Vulture for 1 year without any drinking water. He points out that because of their high body temperature, ease of mobility, carnivorous diet, and perhaps functional salt glands, Turkey Vultures are well-adapted to arid environments.

Birds produce more metabolic water than other vertebrates because they have a higher metabolic rate (Gill 1990). Metabolic water is produced at a rate of 1.07 g (grams) per 1 g fat, 0.56 g per 1 g carbohydrate, and 0.40 g per 1 g protein. Birds also conserve water by excreting a semisolid uric acid rather than a very watery urea as mammals do (Bartholomew and Cade 1963). Birds that are exposed to the diurnal rigors of desert life expend much more water than mammals for evaporative cooling; vultures, however, can easily escape the need for evaporative cooling by soaring at high elevations.

2.3.2
Water Loss Through Evaporation

Birds do not sweat, but they do use body water for cooling through the processes of panting and gular flutter (Calder and Schmidt-Nielsen 1967). Evaporative cooling, resulting in a high rate of water loss, is believed to be characteristic of birds in general (Bartholomew and Cade 1963) and probably depletes water from desert birds more than any other factor. Fortunately for vultures, there is an inverse relationship between body weight and water loss (Bartholomew and Dawson 1953). Even though the amount of water lost by evaporation is still greater than the amount of water produced metabolically, the difference between the two is minimal in larger birds. The additional water required to make up the difference can be obtained from eating meat or carrion, food sources rich in water.

Why waste water to cool your body? The forces of natural selection would certainly choose birds with the least need for evaporative cooling to occupy desert niches. An alternative solution would be to simply have one's body temperature rise as the day gets hotter. This is, in fact, what some desert birds do, within limits, since the enzymes necessary for metabolism only work within a narrow temperature range. Many large birds, like vultures, employ hyperthermia to minimize evaporative water loss and in some cases, at least, can dissipate heat without water. While the environment is warming, a Turkey Vulture will allow its body temperature to rise from a low of 38 °C (100 °F) at 5 a.m. to over 40 °C (104 °F) at 1 p.m. (Hatch 1970). By allowing its body temperature to rise with the rising ambient temperature the Turkey Vulture eliminates the need to waste water with evaporative cooling. As the ambient temperature drops in the evening, the bird dissipates its stored body heat to the rapidly cooling environment. In this way heat is removed by radiation and convection rather than evaporation, and precious water is not sacrificed. In fact, evaporative water loss is probably necessary only under extreme conditions, such as when a vulture is subjected to heated chambers in laboratory experiments, where it cannot escape the thermal stress. A bird in the wild need only soar up away from the hot desert floor to obtain a cool respite.

At cooler times, that is when the ambient temperature is below the zone of thermoneutrality, Turkey Vultures minimize heat and water loss (Arad and Bernstein 1988). Only 13%-22% of their metabolic heat production is lost by water evaporation when the ambient temperature is below 26 °C (79 °F).

CHAPTER 3

Hawks and Allies

The hawk is the quintessential raptor. A hawk's appearance, as a ferocious predator, is accentuated by adaptations for visual foraging that include a bony shelf over the eyes and, in many, black feathers about the eyes, reminiscent of, perhaps, a bandit's mask. The bony shelf may shade the eyes from the sun and wind, or serve as a dust deflector (Johnsgard 1990). Black feathers, in front of and below the eye, reduce glare, like the black smudges that football players (both North American football and "soccer") apply to their faces (R. Guthrie, personal communication).

Of the many forms of hawk-like birds, deserts play host to a restricted subset, including a variety of falcons, some eagles, kites, accipiters, buteos, a caracara, and in grassy borders, the unique Secretarybird (*Sagittarius serpentarius*). These birds are all members of the order Falconiformes, along with the vultures treated in Chap. 2.

3.1 Predatory Behavior in Hawks and Allies

Populations of hawks inhabiting desert regions may, in some cases, be ecologically distinct from the rest of their species. These "ecotypes" may possess subtle behavioral and physiological characteristics that differentiate them from their nondesert conspecifics. Among Ferruginous Hawks (*Buteo regalis*) (Fig. 6) of North America, for example, individuals that nest in deserts also winter in desert habitats and migrate through desert areas (Schmutz and Fyfe 1987). By the same token, Ferruginous Hawks that nest in grassland areas also winter and migrate in grassland areas. The desert Ferruginous Hawks are also distinguished from their grassland counterparts by diet. They eat a much larger percentage of lagomorphs (rabbits, hares, and picas), while grassland inhabitants eat mostly rodents. Although this may simply be a case of taking the food that is most available, Schmutz and Fyfe suggest that hunting tactics for rabbits would probably be different than for rodents, so that the desert Ferruginous Hawks may differ in hunting behavior and morphology, and thereby possess specialized adaptations that would be reflected in a difference in genetic makeup between the desert and grassland populations.

Fig. 6. Desert populations of Ferruginous Hawks (*Buteo regalis*) may have subtile genetic differences associated with feeding primarily on lagomorphs rather than rodents. (Photograph by Tony Leukering)

In West Africa, the greatest diversity of raptor species can be found on the grasslands between the desert and the forest (Thiollay and Clobert 1990). Habitat segregation, even on grasslands, provides a variety of niches. Some species, like the Gray Kestrel (*Falco ardosiacèus*) select the drier, more open areas and may possess adaptations that allow them to range more frequently into desert areas. In the desert itself, limited resources result in each individual raptor maintaining a discrete hunting territory, allowing little or no overlap with other members of its species (Johnsgard 1990).

3.1.1 Euryphagy

The desert menu not only lists relatively few choices, but the quantity of any one selection is limited as well. A desert hawk must, therefore, be willing and able to eat whatever prey items are available. In deserts, even seemingly unpalatable species, such as the heavily armored horned lizards (*Phrynosoma* sp.) are captured and consumed (Miller 1948; Milne and Milne 1950; Bundy et al. 1955; Tanner and Krogh 1973). Swainson's Hawks (*Buteo swainsoni*) flip the lizards over, devour the viscera from the ventral side, and leave the skull armament uneaten (Sherbrooke and Montanucci 1988). As a result of this dietary indiscrimination, a diversified cuisine is a tendency that has been well established among desert hawks. It has been suggested that, as a rule, diurnal birds of prey are opportunistic in the prey that they select, a characteristic that would seem to preadapt them to desert conditions. According to Jaksić and Braker (1983), increased diversity among raptors in a particular habitat is not due to specialization, but to a higher prey abundance that allows more species to feed. These authors contend that raptors take different sized prey according to local availability without regard to considerations such as prey weight. Thus, the diversity of raptor species in deserts may be limited more by the quantity than by the diversity of prey available.

Examples of euryphagy in falconiform birds are provided by Tawny Eagles (*Aquila repax*) and Golden Eagles (*Aquila chrysaetos*). The Tawny Eagles that inhabit open, arid parts of Africa eat anything from termites to dead elephants, and kill anything up to the size of a 4 kg dik-dik (*Madoqua* sp.) (Brown et al. 1982). Golden Eagles, the largest predatory birds of terrestrial habitats in North America, normally eat lagomorphs but will also attack Great Blue Herons (*Ardea herodias*), Whooping Cranes (*Grus americana*), small Canada Geese (*Branta canadensis minima*), and will occasionally attempt to kill a Pronghorn Antelope fawn (*Antilocapra americana*), although that takes considerable effort on the part of the eagle (Goodwin 1977; Windingstad et al. 1981; McWilliams et al. 1994).

A generalized feeder, like the Golden Eagle, has a long digestive tract designed to digest food very efficiently (Barton and Houston 1993). Desert predators typically fall into the category of generalized predators and have a low metabolic rate coupled with a high digestive efficiency. Such raptors are usually "searchers" and are able to eat poorer quality foods than specialist feeders. According to Barton and Houston, poor-quality foods include mammals, earthworms, and carrion. The Crested Caracara (*Polyborus plancus*) is perhaps the least discriminating diner, carrying the concept of euryphagy to extremes, as it eats regurgitated carrion from vultures (Wagner 1980). This poor-quality meal is had at the expense of harassing vultures in flight, causing the vultures to disgorge the contents of their crop, and consuming the ejecta in midair.

Specialist feeders, like Peregrine Falcons (*Falco peregrinus*), in contrast to the generalized feeders, have shorter intestines and need higher-quality foods such as birds with fat. Their food requirements demand that they must be "attackers" and must pursue high-quality prey species. Specialists process their food quickly. Food moves through their alimentary tract and the waste is eliminated rapidly so that little time is lost before the birds are able to return to peak hunting performance. In search of high-quality foods, even other raptors are considered fair game by Peregrine Falcons (Klem et al. 1985). Prey items must have sufficient fat to be high quality. A diet of starved rodents will not, for example, provide sufficient nutrition for an American Kestrel (*Falco sparverius*) (Taylor et al. 1991). Only where a sufficient quantity of high quality food is available will a desert offer a niche to a specialist feeder.

If raptorial specialists are rare in deserts, those that specialize in eating fish might not be expected at all. However, fish-eating birds do sometimes find suitable forage in desert areas. Bald Eagles (*Haliaeetus leucocephalus*) are considered to be specialized as piscivorous raptors even though they are opportunistic, feeding on other items as well. They will consume carrion, mammals, birds, reptiles, and amphibians, but even in the Sonoran Desert of Arizona 76% of their diet consists of fish (Grubb 1995). The Verde and Salt Rivers, flowing through the Sonoran Desert north and east of Phoenix, respectively, provide adequate food to support both a wintering population and a breeding population of Bald Eagles. The breeding birds nest in the riparian vegetation, which includes several species of broadleaf trees such as cottonwoods (*Populus fremontii*), willows (*Salix goodingii*), planes (*Platanus wrightii*), ashes (*Fraxinus pennsylvanica*), and alders (*Alnus oblongifolia*).

The diversified cuisine concept might be a little misleading in some cases. Depending upon the region, a desert hawk might have to be content with a menu supplying a large quantity of one item with very small quantities of a variety of supplemental "condiments". For example, Ferruginous Hawks in some sections of the Great Basin Desert of western North America obtain over 97% of their food bulk from lagomorphs, the other 2-3% being supplied by a variety of other mammals, birds, reptiles, and invertebrates (Bent 1937; Snow 1974). To a lesser extent this is also true of the Ferruginous Hawks of Cedar Valley, Utah, whose major food item (44%) is kangaroo rats (*Dipodomys ordii*), and the Ferruginous Hawks of northeastern Colorado, whose diet is centered (41%) around thirteen-lined ground squirrels (*Citellus tridecemlineatus*). And in central Chile, another species of desert raptor, Harris' Hawks (*Parabuteo unicinctus*) (see Fig. 10), eat primarily diurnal rodents seasoned with a wide variety of other items (Mader 1978; Jaksić et al. 1980).

Seasonal variation in diet is also common among birds (Martin et al. 1961). Swainson's Hawks (Fig. 7) are considered primarily insect eaters as an adaptation to desert resources Johnsgard 1990). However, during the breeding season they take a larger proportion of mice. The nestling Swainson's Hawks presumably require the higher caloric value of vertebrate food for growth.

Fig. 7. Although Swainson's Hawks are primarily insect eaters, as an adaptation to desert resources, young birds, such as this fledgling in Montana, USA, are fed a larger proportion of mice to meet their growth requirements. (Photograph by Mark Vrtiska)

3.1.2 Agility and Diverse Hunting Strategies

A raptor with a predictable food source may be able to survive with a limited set of hunting skills. Osprey (*Pandion haliaetus*), for example, may obtain their entire diet by perching or flying over water and catching fish in their talons. In contrast, a desert raptor must be agile and employ a diverse repertoire of hunting strategies in order to capture a wide variety of food items. Ferruginous Hawks inhabiting deserts (1) hunt from perches, (2) hunt from the ground, (3) course low over the ground, and (4) soar high to search large areas. The time spent on each of these techniques is related to the probability of successful prey capture and the energy cost of the technique itself (flying is more energetically expensive than perching but may increase the probability of locating prey when prey density is low) (Wakeley 1978).

A relatively long tail and short wings make Harris' Hawks agile enough to capture birds as well as mammals (Mader 1978). Coupled with cooperative hunting (see Sect. 3.1.3), the Harris' Hawk is well adapted to gleaning a wide variety of prey items from the desert environment. The rapid pursuit of prey is a dangerous occupation, however, and one environmental factor that selects for the refinement of agility is potential injury from flying into sticks with enough force to be impaled. Evans (1977) indicates that this type of injury, surprisingly perhaps, occurs with some regularity. Yet, there are raptors in deserts, particularly Gray Hawks (*Asturina plagiata* = *Buteo nitidus*) and Cooper's Hawks (*Accipiter cooperii*) of North America, that show amazing skill at hunting in dense brush without injury (Stensrude 1965). Cooper's Hawks, in particular, are morphologically well adapted to hunting in thick desert brush, having short, powerful wings and strong legs (Lamberton 1993).

Some prey species are able to learn which obstacles test the agility of predators and provide effective deterrents against capture. In some places, pigeons have learned to fly under telegraph wires when threatened by Peregrine Falcons (Sutton 1929). The falcons attacking these pigeons would cease their pursuit and veer off when they saw the wire. It is likely that prey species develop a bit of agility themselves, in time, and might even learn to fly very near sharp broken branches.

3.1.3 Cooperative Hunting

Cooperative hunting is practiced by some hawks and falcons and seems to be a desert adaptation where food is scarce (Bednarz 1988b). Groups of two to six Harris' Hawks hunting together are able to capture a wider variety of prey items, including animals that are larger than themselves. The rewards of this coordinated effort are improved capture success and greater caloric intake for each individual hunter.

The cooperative hunting strategies that allow Harris' Hawks to expand their range of prey items are variable and include three observed techniques. (1) Groups of up to six hawks will split up into smaller subgroups of one to three birds and hunt an area in a "leap-frog" fashion. The trailing subgroup, after surveying its area carefully, will fly a short distance in front of the forward subgroup and begin searching the new area. This cooperative hunt will proceed until a prey item is secured and all are able to share in the resulting meal (Bednarz 1988b). (2) If a rabbit is seen dashing for cover under a thick clump of brush, the group of Harris' Hawks will take up positions surrounding the thicket (Fig. 8). One of the hawks will swoop down and enter the brush in pursuit of the rabbit. The rabbit, attempting to escape, will flee from the cover, only to be met, in any direction, by a hawk waiting to pounce on the victim (Gaulden 1993). (3) In yet another example of cooperative hunting, one Harris' Hawk will scare birds out of a tree, then fly off while its companion hides in the tree and waits for the unsuspecting birds to return to their roost.

Ferruginous Hawks inhabiting desert areas also employ cooperative hunting techniques. Two hawks have been observed hunting together, with one flying high while the other flew low over the ground pursuing a rabbit. The

Fig. 8. Harris' Hawks (*Parabuteo unicinctus*) exploit desert resources efficiently by hunting cooperatively. (Illustration by Will Yandik)

rabbit, concentrating on the hawk visible high above, was easily taken by surprise and captured by the low-flying partner (Snow 1974).

In many deserts of the world a cooperative mode of hunting has developed among certain species of falcons (Hector 1986). This particular mode of hunting seems especially tied to desert scrub and desert grassland areas with numerous scattered hiding places. The cooperative-hunting falcons include the Aplomado Falcon (*Falco femoralis*) of Central and South America, the Red-necked Falcon (*F. chicquera*) of Africa and southern Asia, the Brown Falcon (*F. berigora*) of the Australian region, the Laggar Falcon (*F. jugger*) of south central Asia, and the Lanner Falcon (*F. biarmicus*) of Africa and southern Europe.

The hunting strategies of the cooperative-hunting falcons enable them to efficiently capture avian prey. Unlike a solitary Peregrine Falcon, which nearly always captures prey in midair, the cooperative-hunting falcons will also pursue prey on the ground and through trees, using a variety of hunting strategies that can distract and confuse prey. For example, one falcon may hop and run through congested cover in pursuit of birds that are trying to remain concealed, while its partner hovers overhead waiting for a bird to break cover, or the partner may dash into the bush when an unsuspecting bird reveals itself with a sudden movement. Employing another technique, one falcon will fly vertically up into a tree to flush birds up and out so that its partner, hovering over the tree, can effect a midair capture.

3.1.4 Hunting Open Areas

The swift flight of falcons, together with an attending ability to seize a victim, even before detection at times, makes them well adapted to hunting in treeless deserts where concealment is minimal. Some falcons apparently require open areas and even their reproductive success is correlated to available open country for hunting (Varland et al. 1993). In Arizona, North America, a pair of falcons were observed to catch a most challenging prey item, a White-throated Swift (*Aeronautes saxatalis*), in midair. The female falcon chased and distracted the swift while the male stooped and captured the swift from above (Arnold 1942).

One of the more unusual adaptations among desert raptors is that of the Black-shouldered Kite (*Elanus caeruleus*) of the desert island of Masira in the Arabian Sea. The island consists of jagged lava crags in the central highlands, surrounded by low sandy plains, none of which support any rodent populations. Scarcely more than a few lizards occupy the area. Yet, a small population of kites is able to survive and occasionally raise a single nestling by scavenging for scraps of fish in the Arab fishing villages (Grossman and Hamlet 1964).

3.1.5 Hunting from the Ground

Many deserts characteristically lack trees, or even tall cacti, for perches. If a predator wishes to avoid the energy expenditure of hunting from the air, the only alternative is to walk over the ground to stalk prey. This technique is especially useful in desert grasslands where the vegetation can conceal prey items from airborne hunters. The long legs of the Secretarybird are particularly well adapted to this terrestrial mode of foraging. Secretarybirds usually hunt cooperatively, in pairs, walking about a half a mile apart, or just in sight of each other (Grossman and Hamlet 1964). They search for such things as birds, eggs and nestlings in breeding season, snakes, scorpions, amphibians, lizards, rats, and grasshoppers. Sometimes they stomp the ground with their feet to scare up prey items. When a vigorous prey item, like a rodent, is spotted, the Secretarybird will chase the intended victim on foot while beating its wings to distract the prey. The Secretarybird usually captures its prey with its beak, but it may pin the prey to the ground with its feet. Since its beak and feet are relatively weak, larger prey may be dispatched by being carried aloft and dropped to the ground (Storer 1987). However, most of its prey items are small and can be subdued by being beaten against the ground or struck with the feet before being swallowed whole (Brown et al. 1982).

The Pale Chanting Goshawk (*Melierax canorus*), with its long legs, resembles a small Secretarybird, and also captures most of its prey on the ground (Karmali 1980; Brown et al. 1982). The Pale Chanting Goshawk can be found hunting in true desert areas, such as the Namib Desert of western South Africa, where it moves from perch to perch and swoops to the ground to capture prey. If a ground swoop fails, the goshawk may walk about, stomping the ground with its feet like a Secretarybird. Such behavior conjures up images of *Tyrannosaurus rex*, feet thundering over the land with heavy footfalls, the impact tremors frightening prey into jumping up, revealing their positions and exposing them to capture. Paleontological studies which suggest that the ancestors of birds can be found among the dinosaurs (Bakker 1986) make the speculation of such a behavioral similarity plausible.

3.1.6
Hunting at Waterholes

Of the many species attracted to desert waterholes, raptors favor meaty birds like doves, pigeons, and quail as their prey (Cade 1965; Beck et al. 1973). An interesting exception is reported by Arnold (1942), where a pair of falcons seemingly bypassed slow-flying, meaty doves to capture and consume a small mercurial swift. The raptors seldom, if ever, drink at the water holes themselves, indicating that their water needs are, for the most part, met through their carnivorous diet.

Hawks were observed by Beck et al. to visit a Sonoran Desert waterhole primarily at midday, so Gambel's Quail (*Callipepla gambelii*) would come in the morning and evening. One might speculate that the raptors do not visit the waterhole when prey are most abundant because prey are particularly cautious at waterholes, and the many eyes of the quail could detect a raptor soon enough for all to find cover. In that case, during the peak hours of desert activity – morning and evening – raptors might have better hunting success elsewhere. During the midday "doldrums", however, when animal activity drops off dramatically in the desert, raptors may find that waterholes offer the best opportunity for locating any prey at all.

3.1.7 Predator and Prey Conspicuousness

Desert birds of prey tend to be paler than their forest counterparts. Concealment from prey may be one of the selective pressures that account for this phenomenon. It has been noticed that hawks select perch sites that best conceal them from prey. Lighter morphs of Red-tailed Hawks (*Buteo jamaicensis*) perch in more open areas while darker morphs select denser cover (Preston 1980).

The raptors themselves provide a selective pressure favoring the survival of prey that can remain inconspicuous. Although a long-term specific search image is the most important determinant of prey selection, prey made conspicuous by some physical oddity, or movement due to unfamiliarity with their environment, are at greater risk of being detected and captured by a raptor (Mueller 1974, Snyder et al. 1976). So desert predators and prey alike tend to match the coloration of their habitat.

3.2 Breeding Behavior in Hawks and Allies

3.2.1 Breeding Biology Adapted to Predatory Behavior

The Calansho Serir in the Libyan Desert is a vast, arid, gravel plain extending for hundreds of miles between the Calansho Sand Sea to the east and the Tazerbo Sands to the west, and is completely devoid of any vegetation. In July the temperature often reaches 49 °C (120 °F). There is no primary productivity and the area would seem to be unable to support any food chain at all. Yet, Sooty Falcons (*Falco concolor*) live and breed here, under conditions that are considered to be among the harshest on earth (Booth 1961; Fig. 9). The source of food for the nestlings is the autumn migration of birds from Europe (Gaucher et al. 1995). To exploit this resource the Sooty Falcons delay their breeding until August, a rather remarkable adaptation that seems to be unique among Palaearctic birds, except for the Eleanora's Falcons (*Falco eleonorae*) of the Aegean Sea region, which harvest the same migratory birds and also delay their breeding. The Sooty Falcons hunt at dawn and dusk – a time when migrants are often found descending to investigate the desert, or resting

Fig. 9. Sooty Falcons (*Falco concolor*) breed in one of the most arid regions of the Libyan Desert. (Illustration by Will Yandik)

while the temperature is relatively cool and some daylight is available so they can be on the lookout for predators.

At the other end of the African continent, a slightly larger falcon makes annual adjustments to its breeding season in response to changes in food availability. Near Pretoria, South Africa, Greater Kestrels (*Falco rupicoloides*) lay their eggs earlier in the season if the amount of rainfall the previous year was high (Kemp 1991). Greater rainfall in the summer produces healthier vegetation that supports more grasshoppers and small vertebrate prey items for the kestrels. Wellfed female kestrels will lay eggs before the median laying date during the following breeding season. Such early breeding appears to enhance the success of fledging young.

Bednarz (1987) has suggested that a flexible and prolonged breeding season in Harris' Hawks is a favorable adaptation for desert conditions. There are a variety of unpredictable causes of nest failure; foremost among them are violent summer storms and periods of high predation that are frequent in the desert of southeastern New Mexico where Bednarz did his study. The breeding season of Harris' Hawks extends throughout the entire year, an adaptation which allows the birds to renest after a nest failure. They have been observed laying eggs as early as January and fledging young as late as December (Ellis and Whaley 1979). [In Arizona, Mader (1978) documented the Harris' Hawk nesting cycle from February to December, in contrast to Red-tailed Hawks (*Buteo jamaicensis*) whose nesting cycle extended from February to June.] Mader (1977a) witnessed a polyandrous trio of Harris' Hawks lay three clutches of eggs in three separate nests in one year. The first and the third clutches succeeded in producing fledglings but the second failed for unknown reasons. When a second brood is raised by Harris' Hawks in the same year, the first brood becomes independent of the adults during the raising of the second brood. Otherwise, if there is no second brood, the original fledglings will stay with their parents through the winter.

3.2.2 Breeding Season

What is it that makes a prolonged breeding season in Harris' Hawks seem unusual? Or, on the other hand, why do most birds nest at one particular time and not at another?

The eminent ornithologist and systematist Ernst Mayr, in his classic *The Growth of Biological Thought* (1982), points out that there are always two types of answers to questions that ask "why?" in biological systems. The answers fall either into the category of proximate (physiological) causes or ultimate (evolutionary) causes. For example, if one asks the question, "Why does blood flow through the body?", the proximate answer would be that the heart pumps it, while the ultimate answer would be that evolutionary forces have selected this anatomical design to deliver food and oxygen to the cells of the body as well as to transport wastes and other materials through the body.

Likewise, the question, "Why do temperate birds breed in the spring?", has two types of answers. The proximate answer seems to be that birds are responding to the photoperiod, which has increased to a certain minimum number of hours of daylight, and has triggered the physiological changes that initiate breeding behaviors. The ultimate answer, however, is most apt to be that there is generally more food available in the spring to feed nestlings, so evolution has generally selected for those individuals that breed in the spring.

The distribution of Harris' Hawks is primarily neotropical, in regions where food is relatively abundant year-round, and like many tropical species Harris' Hawks may breed at any time of the year, in response to food availability directly, rather than in response to increased photoperiod. In latitudes near the equator, the photoperiod does not vary much anyway. As the range of Harris' Hawks spread northward into the deserts of New Mexico and Arizona, they were not constrained physiologically to breed at any particular season but could postpone their breeding until conditions, such as violent storms, high predation, or other events leading to nest failure, had passed. The best time for breeding in terms of food availability is also unpredictable in the desert, so a prolonged and flexible breeding season allows Harris' Hawks to make temporal adjustments to fit local circumstances. In addition, Mader (1978) has suggested that the prolonged breeding season in Harris' Hawks is made possible by their greater hunting versatility, involving greater agility for the capture of a wider range of prey items as well as cooperative hunting.

It seems that food availability can also affect the sex ratio of fledglings in sexually dimorphic raptors. Among raptor species, females are often considerably larger than the males. When food for nestlings is low, competition may result in one nestling not getting enough food to survive. In a nest where a female hatched first and a male hatched second the size difference between the two nestlings will result in the larger female getting most of the food brought to the nest by the parents, leaving the male to starve to death (Edwards et al. 1988). In other situations, where both nestlings are the same sex, or a male hatches before a female, the size difference is not significant and each nestling would have an equal chance of survival. Edwards et al. assert that Golden Eagles in Idaho, North America, have a male-biased sex ratio of nestlings that is altered towards an even ratio when such periods of decreased food availability occur.

3.2.3 Cooperative Breeding

In desert areas, cover and concealment for nest sites is minimal. Some desert raptors breed in cooperative groups so that there are more than two adults to defend the nest from predators. Only two members of the order Falconiformes regularly breed in cooperative units – Harris' Hawks (Fig. 10) and Galapagos Hawks (*Buteo galapagoensis*). Both species seem to be responding to the ecological constraints imposed by the scarcity of open water sources in the desert, and use nest helpers to improve reproductive success (Dawson and Mannan 1991). The lack of water reduces both plant cover and food supplies, so that nest sites are often exposed to predators, and food for nestlings is difficult to obtain. In order to increase the odds of nestling survival, nests of Harris' Hawks in the Sonoran Desert of southern Arizona are commonly attended by three adults (Mader 1979). Typically two males accompany a single female. Both males share in copulating with the female, incubating the eggs, brooding the young, and feeding the young and the female.

Mader suggests three hypotheses as to why a cooperative breeding system would increase breeding success in a desert environment where food resources are scattered and subject to seasonal fluctuation. (1) An extra adult increases the food supply and thereby insures greater fledgling success. Cooperative hunting also provides more food for the adults as well as the young. (2) A third hunting adult leaves more time for the nest-attending hawk to guard against predators. (3) If one adult dies, the other two can shift their nest duties and still avoid the loss of the brood.

The advantage of extraparental care to guard against predators, in desert situations where cover and concealment is minimal, can even extend beyond species boundaries. In Idaho, a pair of Ferruginous Hawks and a pair of Swainson's Hawks sometimes nest in close proximity, so that there are four adults to defend a nest site that may otherwise be completely exposed and vulnerable to predation (Thurow and White 1983; Woffinden and Murphy 1983). This interspecies cooperation has the advantage that there is less competition for the same food resources: the diet of the Ferruginous Hawks consists of mostly small mammals, while the Swainson's Hawks rely more on birds and insects.

There are a few raptors that will apparently breed in cooperative units only on occasion. Three or more adults of Red-tailed Hawk, Swainson's Hawk, Black-shouldered Kite (*Elanus caeruleus*), and Bateleur (*Terathopius ecaudatus*) have been observed tending a nest (Skutch 1961; Grimes 1976; Santana et al. 1986; Cash 1989). It may be that ecological constraints, such as those imposed by the desert, serve to favor a behavior of cooperation that appears irregularly among raptors as a genetic variant.

An alternative explanation for nest-helping behavior has been proposed. The suggestion is that the feeding of nestlings by birds other than the parents is a byproduct of communal social structure, rather than an adaptation evolved by natural selection (Jamieson 1991; Ligon and Stacy 1991). The helpers may be feeding the young because they share the same instinct to respond to a nestling's demand for food as the parents. In the view of these authors, the social structure of communal nesting may be an adaptive response to ecological constraints, but the feeding of nestlings by nonparents may have

Fig. 10. The Harris' Hawk (*Parabuteo unicinctus*) is one of only two species of hawks that regularly breed in cooperative units. In the Sonoran Desert of southern Arizona, USA three adults commonly attend a nest. (Photograph by Don Roberson)

no unique genetic basis. Opponents of this view feel that these authors are confusing ultimate and proximate causes.

3.2.4 Population Size Regulation

Population levels are lower for raptors in deserts than in other habitats. Desert raptors maintain lower population levels by spacing nest sites at greater distances. The sandy desert soils of the Kalahari in South Africa are less productive than the clay-derived soils of the basalt substrata woodlands bordering the Kalahari (Hustler and Howells 1988). African Hawk-Eagles (*Hieraaetus spilogaster*) breeding on the Kalahari soils space their nests farther apart than the hawk-eagles breeding on the basalt woodland areas. Thus, these habitats exhibit a negative correlation between food production and internest distances.

Let us, for a moment, examine the question of clutch size with an eye toward ultimate and proximate causes. Food supply is probably the most significant critical factor in determining population size in an ecological or evolutionary (ultimate cause) sense. It has been suggested that food supply may also be a proximate cause, and that raptors lay fewer eggs than herbivorous birds because it is harder to obtain meat than plant food (Karasov 1990; Squires et al. 1991). In the Kalahari Desert, Tawny Eagles (*Aquila repax*) produce more single-egg clutches (i.e., fewer two egg clutches) in dry years when food is less available (Hustler and Howells 1989).

There is evidence, however, that food supply is not the only limiting factor in population regulation among desert raptors (Simmons 1993). Simmons suggests that the enhanced survival of healthier fledglings may be a critical factor that selects for reduced numbers of nestlings resulting in lower population numbers.

3.3 Temperature Regulation in Hawks and Allies

3.3.1 Behavioral Adaptations

Desert heat probably has a worse reputation than it deserves. It is unlikely that many indigenous animals actually die because of the heat in a desert, since it is quite easy to find shelter. Insects and lizards, for example, need only burrow a few inches under the sand to enter a cooler habitat. For birds, with the capability of a wide range of movement in three dimensions, escaping the heat is probably even easier. Desert hawks and eagles, like vultures, can avoid much of the midday heat by soaring at altitudes where the temperature is considerably lower than on the desert floor (Fisher et al. 1972). Yet, curiously this assumption remains to be thoroughly tested (Ballam 1984). The act of raising the feathers to "ventilate" the body, or allow air to conduct heat away from the skin, is also a fairly common cooling technique among raptors in warm situations (Palmer 1988b).

Nighttime activity is another way to avoid desert heat, and even though hawks and their allies are typically diurnal, there are a few exceptions where nocturnal activity allows some of these raptors to avoid thermal stress. Harriers, of the genus *Circus*, are able to hunt at night and avoid desert midday heat, because they have evolved owl-like features that permit hunting by sound (Rice 1982). They have a facial disc to pick up sonic cues within 3-4 m. By flying low over fields, at a height of about 2 m, they can find prey by sound alone. Included in this genus is the Black Harrier (*C. maurus*), a bird of arid areas that can be found in the Namib and Kalahari Deserts of South Africa.

Because larger raptors depend upon thermal updrafts to assist them in soaring, falconiform raptors normally migrate during the day. The discovery that Northern Harriers (*Circus cyaneus*) also migrate at night suggests that night migration may occur as a localized strategy to cope with unusual environmental situations such as avoiding desert heat (Russell 1991). In Israel, Levant Sparrowhawks (*Accipiter brevipes*), a smallish species as hawks go, measuring about 35 cm in length, are able to continue their migratory journey at night without the aid of thermals, by using continuous flapping flight (Stark and Liechti 1993).

Since the thermal properties of a nest site can affect nestling survival, choosing a site is an important behavioral adaptation. Nestling falcons, for example, can die from heat stroke (Wilson and Wilson 1983). Prairie Falcons (*Falco mexicanus*) typically nest on cliff faces and seem to select their nest sites on the basis of thermal needs. In hot deserts of New Mexico, Prairie Falcons prefer aeries that face southeast to protect nestlings from the excess heat of midday, while in the cool deserts of North Dakota, preferred aeries face southwest to absorb midday sun and retain warmth against cold nights (Allen 1987).

In one case at least, desert raptors have shown an ability to adapt to artificial (human) modifications in the landscape. In Arizona, Harris' Hawks have moved into suburban areas to cool themselves by bathing in puddles left by backyard sprinklers (Gaulden 1993; Fig. 11). Perhaps ornithologists should be on the lookout for forthcoming adaptations in the form of switching food preferences to cucumber sandwiches and tonic with a twist of lime.

3.3.2 Physiological Adaptations

In terms of thermal regulation, the metabolic production of heat is counterproductive to a desert inhabitant during the day. Raptors from hot environments, such as Harris' Hawk, Tawny Eagle, Black Caracara (*Daptrius ater*), and Black-chested Buzzard-Eagle (*Geranoaetus melanoleucus*) have evolved lower metabolic rates than species from other habitats (Wasser 1986). Furthermore, individual birds can also adjust their resting metabolism to suit environmental conditions such as solar radiation and wind. American Kestrels, Red-tailed Hawks, and Golden Eagles have been shown to reduce their resting metabolism when exposed to an increase in radiant heat or decrease in wind velocity (Hayes and Gessaman 1980).

Even though falconiforms that inhabit desert areas may produce less heat metabolically, they may not have any special adaptations for radiating heat. Wasser (1986) expected desert birds to radiate large amounts of heat from their bodies all the time. However, this did not turn out to be the case. American Kestrels, which are found in a wide range of habitats, are an example. Those inhabiting desert areas do not radiate heat from their bodies any better than nondesert birds. In fact, they have good insulation to conserve heat against cold nights. In order to dissipate heat, desert American Kestrels use the same mechanisms as are available to all members of their species – mechanisms such as panting, gular flutter, evaporating water from facial areas, and allowing heat to be conducted from the legs (Bartholomew and Cade 1957).

Prairie Falcons are a desert adapted species and, unlike other birds of prey, may have some specific physiological adaptations for coping with high temperatures that are not as well developed in nondesert species. For example, they seem better able to produce moisture about the face for evaporative cooling (comparable to sweating, although birds lack sweat glands) and better able to allow their body temperature to rise above normal so that they can conduct heat into a warm environment (Kaiser and Bucher 1985). This increase in body temperature, called hyperthermia, also allows the falcons to endure higher ambient temperatures without resorting to panting and gular flutter, which are costly cooling techniques in terms of water and energy expenditure.

Desert nights are often quite cold because arid air lacks the humidity to store heat from the day. Some raptors allow their body temperature to drop – a physiological response called hypothermia – in order to conserve energy on cold desert nights (Chaplin et al. 1984). The need for insulation on cold desert nights is obvious, as well, and can explain why desert birds do not radiate heat as easily as might be expected of an animal adapted to a hot environment (Wasser 1986). The desert is not always hot.

3.3.3 Morphological Adaptations

The bare tarsus of falconiform raptors allows effective heat exchange, so a lot of body heat can be eliminated via this part of the leg. In warm situations, female Peregrine Falcons at the nest, and nestlings, will rest or sleep with a leg

Fig. 11. Harris' Hawks (*Parabuteo unicinctus*) in Arizona, USA have been observed cooling off in suburban sprinkler puddles. (Illustration by Will Yandik)

exposed or extended back and outward, to dissipate heat and help maintain a stable body temperature (Palmer 1988b). Falcons of warm, dry climates have evolved in such a way that a greater percentage of their tarsus is exposed for heat dissipation (Mosher and White 1978). For example, the ratio between tarsal surface area and body weight in the arid-adapted Lanner Falcon is 1.35, compared to 0.74 for the temperate-adapted Laggar Falcon. The semiarid-adapted Prairie Falcon has an intermediate ratio of 1.08-1.26.

Desert raptors are often paler in color than their counterparts from other habitats. For example, the warmer, drier climate of northern Patagonia in South America seems to select for the pallid form of the Peregrine Falcon (McNutt 1984). The pale coloration of desert raptors may serve more than one function. It may provide better camouflage against the pale, sandy landscape and thus assist in the capture of prey (Preston 1980); it may also assist in thermoregulation, since pale colors reflect solar radiation more than dark colors.

Dark coloration, however, is not necessarily a severe limitation to desert birds. Heat from solar radiation that is absorbed at the surface of the plumage can be easily reradiated and conducted to the surrounding air (Walsberg 1982). Furthermore, the desert is really a thermal "mosaic" and animals can often seek shelter from the heat of the desert floor (Hamilton 1973). So although a pale plumage will reduce heat gain in desert birds, thermoregulation is probably not the major force for the selection of pale coloration in desert birds of prey.

3.4 Water Conservation in Hawks and Allies

3.4.1 Carnivorous Diet

There are three main sources of water available to animals: (1) free water obtained by drinking, (2) food that has water in it, such as meat, and (3) metabolic water released through the process of cellular respiration (the digestive process that converts glucose to carbon dioxide and water, in order to make the energy of food available to cells).

Large predatory birds do not have to drink water to survive, but they will drink free water occasionally when it is available. Cameron (1938), for example, observed eagles (probably Wedge-tailed Eagle *Aquila audax* or Little Eagle *Hieraaetus morphnoides*) and an occasional buzzard (probably Black-breasted Buzzard *Hamirostra melanosternon*) visit a desert waterhole in southwest Queensland, Australia. According to Fisher et al. (1972), the Little Eagle occasionally drinks at waterholes, but there are no records of Wedge-tailed Eagles or Black-breasted Buzzards actually drinking. They come to the waterholes in search of prey. Other raptors that have been observed to occasionally drink at Australian waterholes are Black Kite (*Milvus migrans*), Whistling Eagle (*Haliastur sphenurus*), Brown Goshawk (*Accipiter fasciatus*), Collared Sparrowhawk (*Accipiter cirrocephalus*), and Brown Falcon.

Even though raptors have been observed to drink free water occasionally, it is not necessary for them to drink in order to maintain water balance under moderate conditions (Bartholomew and Cade 1963). They obtain large amounts of water from their food (Bartholomew and Cade 1957; Fisher et al. 1972; Kaiser and Bucher 1985). The fact that raptors in general are not dependent on a reliable source of free water for survival would have facilitated the expansion of the ranges of many species into arid environments. Their water requirements are met by their diet of seed-eating birds and insectivorous birds that make up the bulk of the desert avifauna, as well as the small mammals, reptiles, insects, and other small animals that frequent the desert floor.

3.4.2
Physiology of Water Conservation

In general, falconiforms seem preadapted to desert life, with water conservation abilities in the form of extracting water from urine and feces and secreting excess salt from nasal glands (Cade and Greenwald 1966). Further, there is some evidence that desert birds may be able to absorb more water from feces in the rectum and adjacent areas of the digestive tract than non-desert species (Thomas 1982).

The mechanism of water absorption involves transporting salt from the urine into the blood at the cloaca. This makes the blood hypertonic to (saltier than) the urine. Water will then diffuse passively (by osmosis) out of the cloaca and into the blood, and thus will be kept within the bird's body rather than lost with the urine. The salt that was transported into the blood must then be excreted. Glands in the nostrils are able to remove the salt from the blood by secreting a solution that is much saltier than the blood (Schmidt-Nielsen 1978).

3.4.3
Water Conservation During Breeding

Desert birds of prey lay their eggs at intervals of about a day, like most birds, but unlike many birds, they do not wait until the entire clutch is laid before they start incubating them. Incubation is begun as soon as each egg is laid. Observations of Golden Eagles, Northern Goshawks (*Accipiter gentilis*), Cooper's Hawks (*A. cooperii*), and Sharp-shinned Hawks (*A. striatus*) have shown that chicks will hatch a day or so apart and will vary in size accordingly. The result of this "asynchronous hatching" is that if food supplies (the source of water as well) are low, the smaller chick or chicks will die and allow the older chick or chicks to survive, rather than all of the nestlings being lost (Snyder and Snyder 1979; Edwards et al. 1988; Lamberton 1993). Swainson's Hawk chicks have even been observed to eat their younger siblings when food is scarce (Pilz and Seibert 1978). Water shortages during the breeding season will, therefore, result in a limiting of the population size to a level that can subsist on the available water supply.

CHAPTER 4

ROADRUNNERS

Mark Catesby, the celebrated eighteenthth-century naturalist, illustrator of North American birds and contemporary of John James Audubon, called the roadrunner the Ground Cuckoo. Of all predatory birds inhabiting arid regions, the roadrunner is one of a very few specifically adapted desert forms. The Cream-colored Courser (*Cursorius cursor*) of the order Charadriiformes (shorebirds etc.) is a ground dwelling desert inhabitant of Africa and Asia, and an ecological equivalent to roadrunners.

There are only two species of roadrunners in the world; both are confined to arid regions of the New World. The Greater Roadrunner (*Geococcyx californianus*) occurs in desert scrub, chaparral, and various open arid situations from western United States at roughly 40° northern latitude southward to central Mexico at roughly 20° northern latitude. The Lesser Roadrunner (*Geococcyx velox*) is found in arid semiopen country with tangles and dense shrubby undergrowth from western Mexico at roughly 25° northern latitude southward to central Nicaragua at roughly 13° northern latitude (American Ornithologists' Union 1983).

Roadrunners are members of the order Cuculiformes along with cuckoos and anis. The scientific literature reported here is confined to the Greater Roadrunner. The Lesser Roadrunner is an elusive bird and remains relatively unstudied.

4.1 Predatory Behavior in Roadrunners

4.1.1 Diurnal Activity

Roadrunners are only active during the day (Kavanau and Ramos 1970). Their diurnal habits expose them to the harshest of desert conditions and make them subject to considerable respiratory and cutaneous evaporative water loss. A roadrunner could take a lesson from a kangaroo rat and feed at night. Desert rodents like kangaroo rats (*Microdipodops* and *Dipodomys*) are active primarily at night in low temperatures and spend the day in their burrows, where the humidity is higher than outside. They are thereby able to conserve water by avoiding the daytime desert heat. In fact, these rodents may never drink water throughout their entire lives and get all of the water they need from their diet of seeds and plant tissues. This is so because they produce more metabolic water than they lose by respiration. Roadrunners, on the other hand, are active during the day, are exposed to extreme desert heat, respire air of low humidity, and are frequently faced with the necessity of cooling by the evaporation of water (Bartholomew and Dawson 1953).

Why don't roadrunners hunt at night? Presumably they cannot see, hear, or smell well enough for nocturnal operations. Roadrunners are strictly diurnal. All of their activities occur during daylight. Even during the day, dim light inhibits activity in roadrunners. The visual system of birds other than owls and perhaps a few others, including some goatsuckers and a parrot, is apparently not sufficient to allow nighttime hunting. As humans, we can appreciate this constraint because, unlike that of many fellow mammals, our hunting prowess is essentially nil during hours of darkness. Yet, we must not be blinkered by the mistaken notion that roadrunners have a visual input that is just like our own. On the contrary, recent evidence indicates that birds see in portions of the electromagnetic spectrum that humans do not (Goldsmith 1980; Chen et al. 1984; Parrish et al. 1984). Still, despite their visual acuity, roadrunners do not have the ability to see well at night.

As any mole would attest, however, a highly developed sense of vision is not the only key to nighttime hunting. Some birds like owls and harriers can hunt effectively by hearing (Payne 1971; Knudsen 1981; Rice 1982), but roadrunners do not have the facial and auditory anatomy that these specialists possess. Hunting by smell at night is probably also beyond the roadrunner's ability. Cobb (1960) points out that members of the Cuculiformes have moderately sized olfactory bulbs. Yet, even though roadrunners probably have a reasonably good sense of smell (they might be able to detect a rodent at close range), smell alone would not provide the directional accuracy needed to pinpoint the location of an active kangaroo rat in the dark. The avian sense of smell is an area that needs much study but roadrunners are not among the groups that have been identified as expected to have an exceptionally well-developed olfactory acuity (Ehrlich et al. 1988). So, lacking any ability to seek out prey at night, roadrunners are obliged to hunt by day.

The daily routine of a roadrunner begins with a period of low activity in the morning while the bird spends a lot of time basking in the sun (Kavanau and Ramos 1970). Feeding activity picks up during midday unless the temperature gets too hot (Calder 1968a). A roadrunner will feed actively in midday temperatures of 31-32 °C (88-90 °F) but when the thermometer sizzles up to 41 °C (106 °F) the bird will seek shade and forgo an extended lunch. A peak of feeding activity ensues just before sunset. This may be due to increased availability of insect food at this time, or may also be prompted by the coming hours of darkness and enforced fasting.

4.1.2 Euryphagy

An inexperienced chef should have no qualms about inviting a roadrunner to dinner. Roadrunners appear to eat practically anything. Their omnivorous diet is an adaptation to the harsh desert environment with limited resources (Hils 1985). A roadrunner cannot afford to be choosy about what it eats. The list of foods taken is truly impressive and includes: lizards (including horned "toads"), small snakes, scorpions, tarantulas and other spiders, centipedes, millipedes, mice, cotton rats, small ground-dwelling birds and eggs, young quail, insects, fruit and seeds (Bent 1940; Zimmerman 1970).

Along with a cosmopolitan appetite goes an array of predatory behaviors designed to secure any one of the variety of prey that might appear in the roadrunner's path. The essence of the agility of the roadrunner in its pursuit of such variable prey items is captured eloquently by Arthur C. Bent (1940). "Watch him race across the sand, full speed, after a lizard. Watch him put out a wing, change his course, throw up a tail, change his course again, plunge headlong into a clump of cactus, and emerge, whacking his limp victim on the ground. Watch him jerk a slender snake from the grass, fling it into the air, grasp it by the head or neck, pummel it with his hard mandibles, and gulp it head first. Watch him stalk a grasshopper, slipping quietly forward, making a sudden rush with wings and tail fully spread, frightening the doomed insect into flight, then leaping three or four feet in the air to snatch it flycatcherwise in his long bill....To reach the insects that were in the midst of the bush, he scrambled noisily through the twigs, caught a few, then sprang back to the ground to catch those that had fallen or jumped out.... Its voracity keeps it on the alert for food. To capture a grasshopper one moment, a race-runner lizard the next, and a tarantula the next requires strength, speed, and prowess."

Rikki-tikki-tavi's American counterpart is the roadrunner. For just as the famed mongoose of Rudyard Kipling's Jungle Book is able to dispatch poisonous snakes in India, so too the roadrunner is able to confront and defeat rattlesnakes in the American desert (Wagner 1980). A typical confrontation begins as the rattlesnake coils to strike and the roadrunner balances to counterattack. As the snake lunges forward, the roadrunner strikes the snake's head with its beak. Repeated assaults quickly dispatch the snake which is consumed without delay.

Roadrunners swallow their prey whole. This can sometimes cause problems. A roadrunner may attempt to swallow a rattlesnake that is too large and end up walking about with a snake tail protruding from its mouth. On the other hand, the roadrunner often uses his beak to trim excess nonnutritive limbs from food items. Scorpions are swallowed with gusto after a concentrated attack on the venomous tail spike renders the arachnids defenseless (Bent 1940). Tarantulas are thrown into the air as one by one the legs are flicked off, then the torso is easily gulped. Birds are eaten after wing and tail feathers are removed (Zimmerman 1970).

If dinner is a relatively large or bony item the roadrunner will pulverize it by slamming it repeatedly against the ground or a rock. In this way the skeleton of an animal can be crushed so that the animal will fit more easily down the roadrunner's gullet. The larger the mouse the more it is beaten (Beal and Gillam 1979). Beating breaks the skull, vertebrae, ribs, and, most importantly, the pelvic and pectoral girdles and limb bones of mice. Since the roadrunner swallows its prey items whole, some animals would seem to be well protected from this sort of consumption. The Texas Horned Lizard (*Phrynosoma cornutum*) is armed with stout spines that would seem to make it invincible to roadrunners. However, if offered nothing else on the menu, the roadrunner will apply the centrifugal-slam technique, swinging the reptile around then bringing it down on a hard surface with great force, to process the large lizard (Sherbrooke 1990). The lizard is swallowed upside down so that the spines do not injure vital internal organs, yet injuries to the mouth and digestive tract do commonly occur. If given a choice, the roadrunner will much more readily eat the smaller, less damaging, Round-tailed Horned Lizard (*Phrynosoma modestum*).

These prey-specific behaviors are part of a wide repertoire of activities that allow the roadrunner to survive in arid environments. Pursuing a wide range of food items is an important desert adaptation – euryphagy allows the bird to take advantage of the diverse desert prey species that individually only occur in limited numbers and would be unable to provide a constant food source to a specialized predator.

4.1.3 Camouflage and Terrestrial Hunting

One day a group of White-throated Swifts (Aeronautes saxatalis) were gathering bits of grit at a dried-up creek bed. Suddenly there was a great commotion.

"Where did that monster come from?" cried a terrified swift. He had just had a narrow escape from being snatched out of the air; one of his companions did not escape. "I never even saw him, then suddenly the air was filled with a massive snapping beak and there were blue and red flashes distracting my attention!"

Bent (1940) recorded an observation of a roadrunner lying in wait in a creek bed and suddenly springing into the air to knock down a swift which passed within its reach.

Roadrunners have blue and red patches of skin behind the eyes, but otherwise they are very well camouflaged. They possess a peppered white and metallic blue-black coloration on the back that allows them to blend well with the mottled shadows of the brushland floor (Miller 1988). The profile of the bird is obscured by a streaked neck and back; the light belly has little contrast with the bright Sonoran sky. The cryptic coloration of the roadrunner allows him to remain concealed in such places as under the shrubbery bordering a creek bed, while waiting to pounce on unsuspecting swifts.

Any swift unfortunate enough to need to come to the ground for any reason in the deserts of southwestern North America puts itself in a precarious position. This arena of activity is the domain of the roadrunner. Roadrunners are aptly named. Long powerful legs propel these birds at speeds up to 25–42 km per hour (15-25 m.p.h.) and allow the birds to outrun most prey items, even swift-moving birds and kangaroo rats. Speed may be more important than stealth in an open habitat. Unlike a dense jungle, where a predator, like a jaguar, may be able to get close to its prey by moving slowly and silently, the desert floor is often sparsely vegetated and the successful ground-hunting predator may have to cross a wide open area quickly to secure a meal.

One foot above the desert floor there may be a dense thicket that creates a canopy as visually obscuring as the canopy above any forest. Whereas a Broad-winged Hawk or Sharp-shinned Hawk can hunt under the canopy and survey the floor of a deciduous forest from 20-30 ft up, the roadrunner must often scan the desert floor from near ground level. The best way to hunt this terrain is to cruise the area on foot. Strong legs and quick agility are important adaptations that allow roadrunners to effectively search out and capture prey in desert thickets.

4.1.4 Prey Selection

Although desert birds of prey have a reputation for eating anything that they can get their beaks on, they sometimes have to make choices. What is it that determines which prey a predator will select if alternatives are available? The reports from the many researchers that have investigated this question seem complex and often contradictory (Mueller 1974; Hils 1985). The prey that offers the most nutrition with the least expenditure of energy, or least likelihood of danger, should be preferred. Roadrunners seem to prefer: (1) large prey items that supply more caloric value than smaller prey; (2) active prey, which might indicate good health of the victim; and (3) inconspicuous prey that may fit a specific search image. Inconspicuousness may also indicate that the prey item is nontoxic (Hamilton 1973; Hils 1985).

4.2 Temperature Regulation in Roadrunners

4.2.1 Adaptations for Warming

A cactus doesn't offer much shade. Arid air is bereft of clouds. Although all life on earth is dependent upon the energy provided by the photons of solar energy emanating from our nearest star, a desert animal has a special relationship with the sun. The roadrunner has adopted a combination of regulatory devices to accommodate the extreme daily temperature fluctuations of the desert, for indeed, the frigid desert nights offer a challenge every bit as demanding as the sweltering midday heat. The roadrunner eagerly awaits the warming rays of the morning sun and takes advantage of their radiant properties with feathers erect exposing extensive areas of black skin (Fig. 12) as well as black feathers on the back (Ohmart and Lasiewski 1971). Operating like the black heat-absorbing surfaces of the solar panels that heat our modern solar homes, this darkened skin transforms light energy into heat energy and helps elevate the roadrunner's body temperature from a nighttime low of about 34.3 °C (93.7 °F) to its normal range near 38.4 °C (101.1 °F).

The abundance of black animals, especially poikilotherms, in deserts is a well documented phenomenon and an interesting paradox since excessive heat is a major problem in hot deserts (Hamilton 1973). One would have thought that black coloration, which increases solar heating, would be a disadvantage for a desert animal. Yet, as Buxton (1923) points out, if a desert animal is not the color of its surroundings, it is almost certainly black. Bright greens, reds, blues, and yellows are exceedingly rare in desert fauna. Hamilton asserts that the numerous examples of black coloration in desert animals are an adaptation to enhance radiative heat gain so that the animal can warm up quickly in the early morning sun and then make maximum use of the daylight hours available before the ambient temperature reaches lethal levels. Roadrunners have a distinct advantage in being able to expose their black skin and plumage to the morning sun, then conceal it to retard heat gain as the ambient temperature rises, and at the same time they acquire the camouflage advantage provided by their lighter outer plumage.

4.2.1.1 Sunbathing

Is that bird hurt or does it merely aspire to become a punk rock star? One might ask just such a question when first encountering a roadrunner in its sunbathing posture (Fig. 13). In order to allow the sun's rays to reach his dark interscapular skin, the roadrunner erects his mantle feathers in a most unbirdlike fashion. Rather than producing a smooth aerodynamic surface

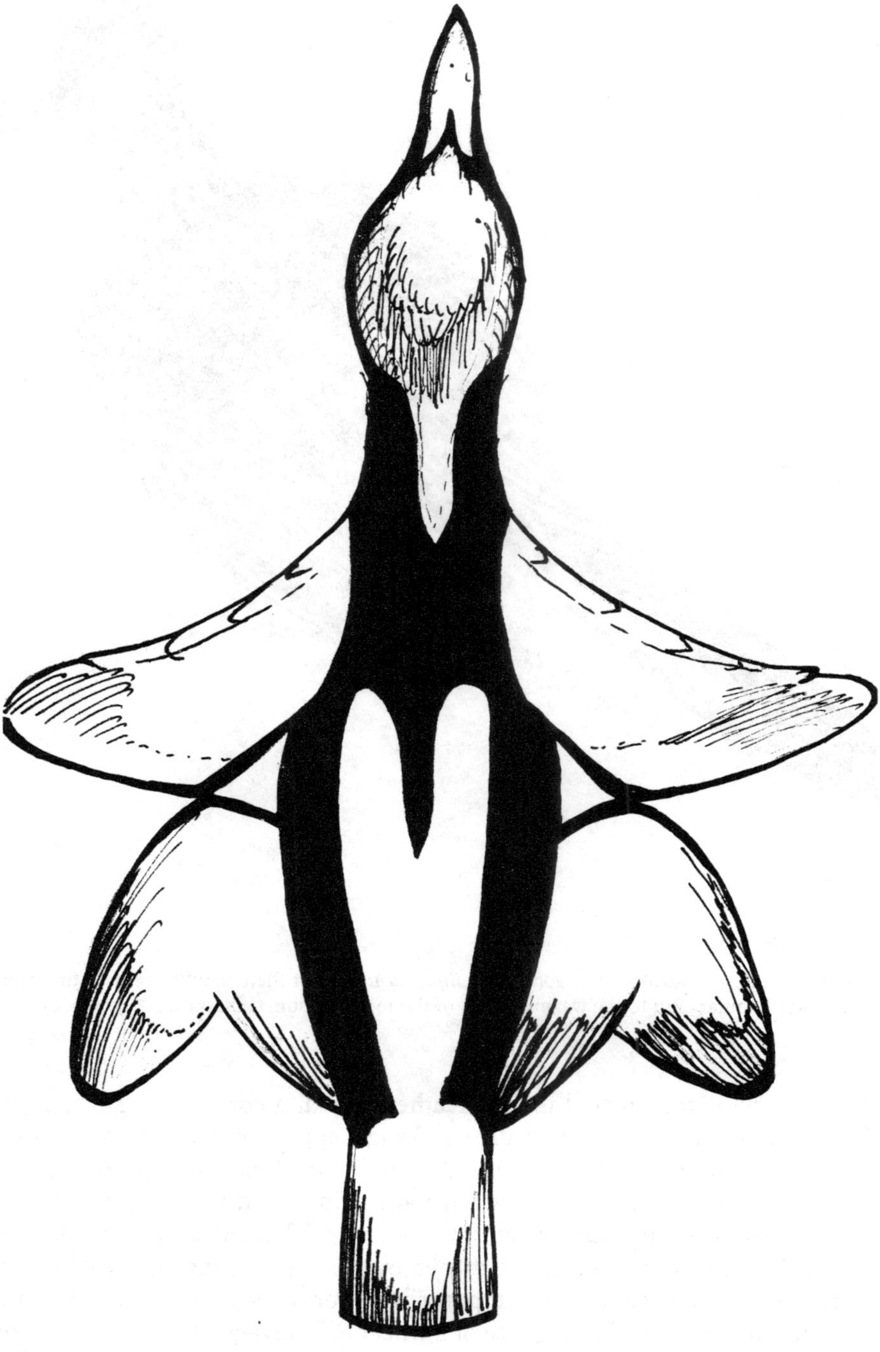

Fig. 12. Dark pigmented skin on the back of the Greater Roadrunner (*Geococcyx californianus*) helps regulate body temperature and conserve metabolic energy by absorbing solar radiation. (Illustration by Will Yandik)

Fig. 13. Greater Roadrunners (*Geococcyx californianus*) erect their mantle feathers to expose dark interscapular skin to the warming rays of the morning sun. (Illustration © James Coe)

over the back, the erected mantle feathers reveal a comical tangle of fluffy black semiplumes creating a rather unkempt appearance, but also revealing the dark feathers and skin that can absorb the much desired heat of the sun. At this time the wings, which are normally held over the back, are moved to the side so as not to interfere with the action of this particular heat sink. If the bird did not appear to thoroughly enjoy this sun basking, one would surely suspect that he was at least temporarily out of sorts, if not in fact critically injured. He has all of the elegance of a wheat field ravaged by a hail storm, an unsettling sight that imparts a sense of damage, certainly not the image of an unadulterated bird. One could scarcely believe that a bird could purposely inflict such disarray on its own coiffure, except that the obvious contentment of the bird mitigates any real sense of danger.

4.2.1.2
Hypothermia

The sun is perhaps the most consistent source of energy available on the desert floor. While food for metabolic energy is scarce and highly variable, both in terms of type and abundance, a cloudy day is rare. The sun provides a dependable reprieve from the chilling cold of the desert night. For without the blanket of moisture laden clouds to retain heat at night, the desert air loses its heat rapidly. Once the ground has radiated its stored heat, the temperature of the desert floor drops dramatically. When the ambient temperature falls below 10 °C (50 °F) the roadrunner compensates by reducing its body temperature. This attainment of hypothermia for the conservation of energy, comparable to torpor in smaller mammals and birds, is an advantage in the nutrient-poor desert environment. The roadrunner then does not have to expend so much metabolic energy to maintain survival through the desert night. The ability to tolerate hypothermia, and the behavior of sunning, are used by the roadrunner throughout the year, but their importance is undoubtedly most critical during the winter months. During the winter desert productivity drops. Although the temperature does indeed go down, it is the reduced food availability, especially in terms of insect numbers, making metabolic energy difficult to obtain, that puts the crunch on the roadrunner's energy budget. Hypothermia and sunbathing in the roadrunner are adaptations that help meet the challenge of cold, dry desert nights and winters.

"Don't you dare let those eggs get cold, Honey!" she said as she drifted off to sleep and allowed herself the luxury of hypothermic energy conservation. An interesting exception to cold night hypothermic energy conservation in roadrunners is the situation in which the incubation of eggs requires warmth. Roadrunners are one of the few bird species that place this task primarily on the male, especially at night (Vehrencamp 1982). Therefore, even though it is very costly in terms of energy expenditure, male roadrunners maintain normal body temperature at night when incubating.

4.2.2
Adaptations for Cooling

Blistering midday heat. The morning is past. An active period of feeding from 7 a.m. to anywhere between 11 a.m. and 1 p.m. (Ohmart 1972) has left the palate sated – for the time being at least. The desert sand is baking and ambient air temperatures may reach 49 °C (120 °F) or more. Although the roadrunner, among birds, may be particularly reliant on the sun and accustomed to its radiant power, the laws of maximums and minimums set levels of tolerance for all physical parameters of life, and roadrunners can withstand only so much heat. Birds perish when their body temperature exceeds 46-48 °C (115-118 °F) (Brook and Birkhead 1991). The roadrunner is, however, especially well suited to the rigors of desert heat. Although a roadrunner will never saturate its University of Arizona T-shirt with perspiration, since avian evolution never reinvented the mammalian sweat gland, the roadrunner does employ a variety of cooling mechanisms.

4.2.2.1 *Legs as Thermoregulators*

The legs of birds have been shown to be important in avian thermoregulation (Steen and Steen 1965). In hot weather blood is pumped quickly along the surface of the legs so that the maximum amount of heat can be released to the surrounding air. This is in contrast to the cold-weather system that shunts blood through a special network of vessels where the blood flows slowly, and arterial heat is conserved by transferring it to venous blood returning to the body.

4.2.2.2 *Hyperthermia*

Roadrunners also use hyperthermia (allowing the body temperature to rise) to contend with midday summer heat. As the ambient temperature approaches normal body temperature (38.4 °C, 101.1 °F), desert birds allow their body temperature to rise in order to maintain a favorable temperature gradient for heat loss via conduction and radiation. This mechanism can continue to the bird's body temperature limit of 46-48 °C (115-118 °F), but once the ambient temperature reaches or exceeds body temperature the only avenue of heat loss remaining is by the evaporation of water. Birds have at least three ways to use body water for cooling: panting, gular flutter, and evaporation from the surface of the skin.

4.2.2.3 *Panting*

Panting involves rapid breathing movements of the thorax and abdominal muscles that move air across respiratory surface areas (Calder and Schmidt-Nielsen 1967). Panting therefore increases the amount of air that passes over moist respiratory surfaces and boosts evaporative cooling. The respiratory anatomy of birds differs markedly from that of mammals and involves air sacs into which air may move apparently either by going through the lungs or by taking bronchial routes that avoid the lungs (see Fig. 2). Since panting is designed to cool the animal rather than increase respiratory gas exchange, there is evidence that air flow during panting avoids lung tissue at ambient temperatures below 40 °C (104 °F). The evidence referred to is the fact that there is no rise in blood pH during panting at these temperatures (Calder and Schmidt-Nielsen 1966). An increase in gas exchange would remove larger quantities of carbon dioxide from the blood. This would increase the blood pH, that is, make it more basic (carbon dioxide lowers the pH or acidifies blood by forming a weak acid, carbonic acid). Thus, the observation that respiratory alkalosis does not occur during panting in roadrunners and pigeons at ambient temperatures up to 40 °C (104 °F) suggests that panting does not increase gas exchange, and that the respired air is probably being shunted around the lung tissue.

If panting did cause respiratory alkalosis, the behavior could cause physiological problems, and not be particularly helpful to the bird. In humans, "panting" or hyperventilation causes respiratory alkalosis with a host of characteristics referred to as hyperventilation syndrome: (1) tinnitus, or ringing in the ear, (2) palpitations, or rapid throbbing of the heart perceptible to the patient, (3) sweating, (4) dry mouth, (5) trembling, (6) pain in the chest and upper abdomen, (7) nausea and vomiting, (8) blurred vision, (9) convulsions and loss of consciousness (Taylor et al. 1989). Without trying to sort out the differences between human and avian responses to respiratory alkalosis, it is probably safe to assume that there would be a negative impact on the bird's physiology. If, for example, vomiting occurred it would be an uncalled for loss of water, and a dry mouth might retard the evaporative cooling process itself. It seems important, therefore, that, if birds are going to use their moist respiratory tract for cooling, they do it without excess gas exchange in the lungs that would lead to respiratory alkalosis.

There is some evidence that panting may cause respiratory alkalosis under certain circumstances. At temperatures above 45 °C (113 °F), under laboratory conditions, respiration in pigeons is increased, carbon dioxide is removed from the blood, and alkalosis does occur (Calder and Schmidt-Nielsen 1966). If higher temperatures can cause respiratory alkalosis in roadrunners it would be to the bird's advantage to keep its body temperature below such extreme levels by using other behavioral and physiological techniques. The roadrunner does, in fact, have other weapons in its arsenal of temperature regulation behaviors.

4.2.2.4
Gular Flutter

Roadrunners can also cool off by evaporating water from their mouth and throat, in a process called gular flutter. Gular flutter is accomplished by a rapid movement of the thin floor of the buccal cavity (interior of the mouth) and the upper throat region. In this manner evaporation of water from the moist tissues of the mouth and throat allows cooling of the blood. It is assumed that capillary dilation in the buccal cavity raises the vapor pressure of the evaporative surface area to enhance evaporation and heat dissipation (Salt 1964). Furthermore, it is believed that the water content of the moisture covering the membranes of the buccal region can be varied to modify the vapor pressure of the evaporative surface area, and thereby enhance heat dissipation during heat stress.

The buccal region is particularly well developed in roadrunners, and may account for their ability to achieve respiratory cooling with a considerably lower energy output than other birds. Calder and Schmidt-Nielsen (1967) have suggested that gular flutter, consisting of oscillations of the buccal surfaces, requires less energy than the movements of the thorax and abdomen needed for panting. This, coupled with the fact that the gular proportions of the roadrunner are much larger than those of other birds of similar weight, indicates that roadrunners are especially well adapted for evaporative heat loss from the respiratory system at minimal energy expenditure. Calder (1968a) has even seen a roadrunner, in heat stress, save yet more energy by gaping in a strong wind, so that it did not need to exercise its gular flutter mechanism to move air across its buccal surface area.

4.2.2.5 Evaporation from the Skin

Because birds lack sweat glands, evaporation from the skin was never considered an important source of cooling until Lasiewski et al. (1971) demonstrated that 45% or more of a bird's evaporative water loss is through the skin at 30-35 °C (86-95 °F). These authors feel that "cutaneous evaporation may represent a major avenue of heat dissipation in many species of birds." The ability of the roadrunner to evaporate water from the skin seems to be about the same as that of other species studied, both desert and nondesert inhabitants, so it is believed that evaporative water loss in the roadrunner is regulated for desert conditions by behavioral rather than physiological techniques. These techniques include standing in the shade of a tree such as a mesquite (*Prosopis* sp.), and holding the wings away from the body to expose the lightly feathered axillary areas to a cooling breeze (Calder 1968a). The sides of many birds are thinly feathered beneath the wings, and are considered important sites of heat dissipation (Schmidt-Nielsen 1978).

4.2.2.6 Rete Mirabile Ophthalmicum

"Having one's brain fried" is a common whimsical comment among my students, who often find the rigors of academic work overtaxing. In the hot desert, however, overheated brain tissue may be a threat to survival. Roadrunners, as well as several other bird species, have been shown to possess a countercurrent blood system to keep brain tissue cooler than the rest of the body during heat stress (Kilgore et al. 1976). This system, called the rete mirabile ophthalmicum, is an elaborate network of blood vessels designed to transfer the warmth from arterial blood flowing toward the brain to cool venous blood returning from the anterior facial region and nasopharyngeal cavities. In this way, the blood entering the brain is cooled and keeps the brain tissues 1-2 °C (2-4 °F) cooler than the rest of the body at body temperatures between 40 and 44 °C (104-111 °F). Although the value of keeping the brain temperature down is uncertain, the advantages of permitting hyperthermia in the body in general while keeping a "cool head" are at least twofold in small birds such as the roadrunner: (1) heat loss by conduction and radiation is enhanced when the body is still warmer than the environment [this occurs at body temperatures of 38.4-46 °C (101.1-115 °F)], and (2) heat gain is reduced as hyperthermia reduces the gap between body and air temperatures, when the air temperature rises above body temperature.

Of the five species of birds studied by Kilgore et al., the roadrunner appears to have the largest rete mirabile ophthalmicum, yet the difference between brain and cloacal temperature of roadrunners was the least of the five. The efficiency of the rete mirabile ophthalmicum depends upon factors other than size, however, so its appearance can easily be misleading. Other factors that influence the efficiency of the rete are: (1) the temperature difference between arterial and venous blood in the rete, (2) the volume of blood flow in the rete, (3) the velocity of blood flow in the rete, and (4) the area and closeness of contact between arterial and venous components of the rete. Nevertheless, at any given cloacal temperature between 40 and 44 °C (104-111 °F), the roadrunner's brain temperature was higher than that of the other four (nondesert) species. This suggests the possibility that the roadrunner's brain may be able to tolerate a higher temperature than nondesert species, and that less energy needs to be expended by the roadrunner to cool its brain – a valuable adaptation indeed, if it turns out to be the case.

As we have seen, roadrunners have a wide variety of cooling mechanisms available. To get the maximum effect from these cooling mechanisms, they will greatly reduce their level of activity during the hottest hours of a summer day (Kavanau and Ramos 1970). The added body heat generated by physical exertion must be avoided at these critical times of severe environmental heat stress. During the heat of the day roadrunners are apt to be found lounging under a mesquite or other shady spot, where they can wait in relative comfort for the sun to go down. Late in the day they return to feeding, and feast on the insects that become active just before sunset. On cool winter days [31-32 °C (88-90 °F)] roadrunners will remain active during midday. So the daily and seasonal activity of roadrunners combines with their physiology to allow them to survive in one of the earth's harshest habitats, the desert floor.

4.3 Water Conservation in Roadrunners

4.3.1 Sources of Water

Water is essential for the metabolic needs of all life on earth, yet the need for water in desert homeotherms is compounded by the desert environment itself. It is a dry environment that quickly evaporates free water, and it also requires water of its homeothermic inhabitants for cooling processes.

Roadrunners obtain water in three ways: (1) the metabolism of food, (2) from the water content of their food, and (3) from drinking free water if available. In the metabolic process that converts food to energy, called cellular respiration, food, which is composed largely of the elements carbon and hydrogen, is broken apart and combined with oxygen to be converted into the two major waste products, carbon dioxide and water. The metabolism of 1 g fat, in addition to releasing 38.5 kJ (kilojoules) of energy, produces 1.07 g water. Likewise the metabolism of 1 g carbohydrate produces 0.56 g water, and the metabolism of 1 g protein produces 0.40 g water (Gill 1990). Birds, because of their high metabolic rate, produce more water than do most vertebrates of comparable body weight.

There is no denying that a rare steak is juicy. A predator's diet of meat supplies a lot of moisture. Even though a roadrunner will drink when water is available, the bird is quite capable of survival on an animal diet without free water (Ohmart et al. 1970). The six roadrunners studied by Ohmart et al. consumed an average of 28.7 g mouse tissue each, per day, under simulated desert conditions with no free water available. The water content of the mouse tissue was estimated to be 66.67%, so each bird obtained about 19.1 ml (milliliters) water from its food each day. These authors also determined that the birds each lost about 26.2 ml water per day, and that the other 7.1 ml were supplied by the metabolic water produced from the digestion of the mouse tissue. So it appears that without any drinking water at all, roadrunners can supply their needs from a diet of meat, with roughly three-quarters of the water coming directly from the animal tissue itself, and the other one-quarter produced indirectly from the metabolism of the tissue.

4.3.2 Water Loss Through Evaporation

Homeothermic organisms like birds expend considerable energy in maintaining a body temperature within fairly narrow limits. When ambient temperatures begin to rise, birds can dissipate excess heat by radiation and conduction as long as the air temperature is lower than the body temperature, but as soon as the air temperature equals or exceeds body temperature, the only avenue of heat dissipation is through the evaporation of water (Salt 1964). Birds of hot deserts are therefore especially vulnerable to respiratory and cutaneous evaporative water loss. Roadrunners, as a consequence of their diurnal lifestyle, must endure the thermal regulatory demands of the most extreme desert heat, and the process of cooling by the evaporation of water is one of the major mechanisms at their disposal. About one-half of the evaporative water loss of a roadrunner occurs through the skin (Lasiewski et al. 1971). The rest of the evaporative water loss is from the respiratory system. Their cutaneous water loss seems to be as great as in nondesert species, according to Lasiewski et al.; no special ability to retard water loss from the skin could be demonstrated in xerophilic roadrunners compared to mesophilic species, and desert birds show no special adaptations for the prevention of respiratory water loss either (Bartholomew and Dawson 1953). So even though birds produce more metabolic water than other vertebrates, they lose more water by evaporation than they produce metabolically. This fact alone prevents most birds from inhabiting deserts.

4.3.3 Adaptations for Water Conservation

Roadrunners cannot afford to waste water if they are to survive on the desert floor. Their only source of water, besides metabolic water, may be their food, and they have a considerable need for water for evaporative temperature regulation. They have met this water conservation challenge with a variety of physiological and morphological adaptations: (1) they excrete less water in their urine than many other birds, (2) they absorb water from their feces, (3) they use less water to excrete salt than most birds, (4) they recycle evaporated water in their respiratory system, and (5) they consume the fecal sacs of their nestlings. The following five subsections of this chapter detail these water conservation adaptations.

4.3.3.1 Water Conservation in the Kidney

The avian kidney loses substantially less water than the mammalian kidney in the excretion of nitrogenous wastes. Birds produce uric acid, a more solid form of nitrogenous waste than urea, the form of nitrogen waste that mammals typically excrete (Bartholomew and Cade 1963). Birds use only 0.5-1.0 ml water to excrete 370 mg (milligrams) nitrogen in the form of uric acid, while a mammal uses 20 ml water to excrete the same amount of nitrogen waste in the form of urea (Gill 1990). The semisolid uric acid is moved through the ureters with peristaltic contractions of the smooth muscles around the ureters. In humans, the presence of uric acid, in the form of a kidney stone, makes these peristaltic contractions painfully obvious, but the elimination of nitrogenous wastes in the form of uric acid is not considered normal physiological operation in mammals. So in terms of the normal excretion of nitrogenous wastes, with the formation of a semisolid urine, kidney function in birds conserves water better than its equivalent in mammals.

The same is not true in the case of salt excretion. The mammalian kidney, with a better-developed loop of Henle, is able to produce saltier urine than the avian kidney. The urine of roadrunners is seldom more than 2 times saltier than blood, whereas kangaroo rats tend to make urine that is about 17 times saltier than blood (Calder and Bentley 1967; Schmidt-Nielsen 1978).

Roadrunners, however, tend to produce saltier urine than other birds. Some birds, like the American White Pelican (*Pelecanus erythrorhynchos*), produce a watery urine when free water is available, but produce a salty urine in order to conserve water when it is not. Roadrunners, however, produce salty urine even when they have free water available. The urine of roadrunners has been reported as being hypertonic to, or saltier [453 mOsmol/l (milli Osmoles/liter)] than blood (300-400 mOsmol/l), even when the roadrunner has an unlimited supply of water. In contrast, the urine of other birds becomes hypotonic (231 mOsmol/l in White Pelicans), or less salty than blood, when freshwater is available (Calder and Bentley 1967). In these experiments, the salt concentration of the urine was determined after the urine had passed through the cloaca and had been voided by the bird. As we shall see, the digestive system has the ability to modify the salt and water concentration of the urine. The hypertonic characteristic of roadrunner urine may therefore be, in part, a result of rectal, rather than kidney, function.

4.3.3.2
Water Conservation in the Digestive System

Even after the kidney has finished making a concentrated urine, the process of water conservation is not complete. Unlike mammals, birds do not dump their urine directly. Instead, the urine enters the cloaca, the end of the digestive tract, where it is further modified. Ohmart et al. (1970) have shown that, in roadrunners, urine moves from the cloaca, back up the digestive tract, to the rectum and ceca, where more water can be conserved by being absorbed through the epithelial lining of these structures. This ability to absorb water in the posterior portion of the digestive system (referred to as the integrative segment) seems to be most highly developed in desert species (Amanova 1975; Thomas 1982). The absorption of a large portion of the luminal water produces dry feces in roadrunners (Ohmart et al. 1970b), and the water content of the feces has been shown to be even lower in simulated desert conditions (Ohmart 1972).

With the reabsorption of water in the cloaca, rectum, and cecum, the roadrunner's digestive system is able to concentrate nitrogenous wastes at very high levels. Birds can concentrate nitrogenous wastes in their urine at up to

3000 times the concentration found in their blood, while the most efficient mammals can only concentrate nitrogen in the urine up to 20 or 30 times the concentration of their blood (Gill 1990).

The roadrunner's ability to conserve water goes into high gear under desert conditions when no free water is available. With an unlimited water supply a roadrunner will expend 41.8 ml water per day and replace one-half of its body water in 3.91 days (Ohmart et al. 1970a); under desert conditions, with no water available, and fed a diet of mice, the roadrunner will expend 26.2 ml water per day, and replace one half of its body water in 5.84 days. So when water is not available the roadrunner's water conservation processes become about 1.6 times more efficient.

The mechanism by which water is conserved in the digestive system has been suggested by Schmidt-Nielsen (1978); it involves a technique that utilizes osmosis and a special salt secreting gland. The cloaca transports salt out of the urine and into the blood. This makes the blood hypertonic to the urine, causing water to diffuse passively out of the cloaca and into the blood, and thus the cloacal water is conserved. The excess salt is then removed by the nasal gland.

4.3.3.3
Salt Excretion

The function of the nasal gland to excrete salt was first discovered in seabirds (Schmidt-Nielsen et al. 1958), later in hawks (Cade and Greenwald 1966), and subsequently in roadrunners (Ohmart 1972). The nasal gland becomes especially useful as blood salinity rises with the loss of water through panting, gular flutter, and cutaneous evaporation. The excretion of salt counterbalances the loss of water to maintain blood salinity within the normal range of 300-400 mOsmol/l. The fluid secreted by the roadrunner's salt gland is about five times as salty as blood (averaging about 1600 mOsmol/l).

The nasal gland is especially important in the rearing of young, as it allows parents to leave the young at critical feeding times. A primary food item for nestlings is the whip-tailed lizard (*Cnemidophorus* sp.) which is active from 7 a.m. to 1 p.m.. Adults must hunt at this time, and leave the nestlings while the temperature is still cool. The black-skinned nestlings warm up by absorbing solar energy. Later, they cool off by gular flutter, with the associated evaporative water loss, increased salinity of the blood to conserve cloacal water, and nasal salt secretion. Parents will regurgitate clear liquid to the nestlings at each feeding to replace water lost, then shade the nestlings during the hottest part of the day. The secretion of salt by the nestlings is therefore an integral part of the synchronization of physiological and predator-prey activities of the species (Ohmart 1972).

4.3.3.4
Water Conservation in the Respiratory System

Desert air is dry. In Frank Herbert's science fiction novel, *Dune,* and its sequels, the author incorporates a device called the still suit to conserve and recycle human respiratory water vapor for desert survival. Indeed, the respiratory system is a potential site of dramatic water loss via evaporation. This is especially true in roadrunners, which rely on the cooling effect of evaporation from respiratory surfaces in the processes of panting and gular flutter (see Sects. 4.2.2.3 and 4.2.2.4). Fortunately, a countercurrent system is able to conserve much of the water used in respiratory cooling in birds (Gill 1990). As dry air is inhaled, evaporation of water from respiratory surface areas cools the respiratory passages. The warm moist air that is then exhaled from the lungs is cooled as it passes through the respiratory passages. The moist air is cooled, and the water vapor condenses on the nasal passages before a now drier air is exhaled from the nares. The conserved moisture is then available for reevaporation on the next inhalation, to continue the cooling process.

So, even though roadrunners, like all birds, evaporate more water than they produce metabolically, they can, by the combined techniques of producing a semisolid urine, excreting dry feces, removing salt in a highly concentrated form from the nasal gland, recycling respiratory cooling water, and consuming food high in water content, effectively conserve water and survive in the arid desert environments of North and Central America without drinking free water.

4.3.3.5
Water Conservation in the Reproductive Cycle

When we read in our environmental science literature about Kansas residents drinking purified effluent during periods of drought, we may shudder at the thought, and take solace in the hope that technology can be depended upon to maintain the purity of this last resort in terms of water sources. It seems unnatural to consume one's own waste products, yet a peculiar situation occurs in roadrunners where parents are obliged to consume the liquid waste products of their young. Parents often have to contend with the fact that children are inherently wasteful. Roadrunner chicks cannot concentrate urine as effectively as adults, so the adults ingest the fecal sacs of their nestlings, reprocess the wastes more efficiently, and thereby conserve a small quantity of water (Calder 1968b). Parental sacrifice of dignity knows no bounds.

CHAPTER 5

Owls

The Owl and the Pussy-Cat went to sea
In a beautiful pea-green boat...

wrote Edward Lear in the mid-nineteenth century. But I never understood what an owl and a cat could have in common to make them consider each other compatible. Perhaps the fact that both are predators would give them the advantage of not being in conflict about what foods to procure during their voyage. Lear's lyric, as we know, portrays a romantic liaison, the owl being, as one might infer from his work, a caricature of Lear himself. But in my imagination they are a Hume's Owl (*Strix butleri*) and a Sand Cat (*Felis margarita*) escaping from a British zoo, rounding Gibraltar, and bound for the east coast of the Mediterranean to return to their desert home in the Middle East.

As attested by Lear's nonsense, owls have inspired the imagination of humans more than any other group of birds. Whether it is due to their anthropomorphic face or to their eerie nocturnal vocalizations, owls have become a part of countless myths, legends, and works of literature. Unfortunately, since owls are difficult to study, much remains to be learned, and little of this literature provides information about desert adaptations. Although considerable attention has been given to the special attributes of hearing and seeing for nocturnal hunting in owls there, are relatively few studies of physiological adaptations in desert owls.

Taxonomists have distinguished two families of owls in the world, Barn Owls and a few related species being placed in a family (Tytonidae) apart from all of the rest of the owls (Strigidae). Anatomical features have been used to establish a phylogenetic distinction between these two groups, however, they seem very similar on a molecular level. Recently, a successful hybridization between a Barn Owl (*Tyto alba*) and an owl from the Strigidae (Striped Owl *Asio clamator*) provides strong evidence that these family distinctions are probably not justified (De La Torre 1990). Further, there does not seem to be a major difference in their ecological adaptations. The range of behavior and ecological adaptations are a bit broader in the Strigidae simply because this family contains more species and they are spread over a broader range of habitats (consider the fish-eating owls of Asia and Africa as perhaps the most striking example of niche expansion in the Strigidae), but both families are represented in desert locations.

Of all the owls of the world [Sibley and Monroe (1990) list 176 species] only about three species appear to be confined to a desert niche. The strictly desert owls include: (1) the Pallid or Striated Scops-Owl (*Otus brucei*) inhabiting a range beginning at the eastern edge of the Arabian Desert and extending across the Iranian, Thar, and into the Takla-Makan Desert of Mongolia; (2) the Pharaoh Eagle-Owl (*Bubo ascalaphus*) of the Sahara Desert and western Arabian Desert, a species that was, until recently, considered a race of the Eurasian Eagle-Owl (*Bubo bubo*); and (3) Hume's Owl (Fig. 14) of the Arabian Desert. One race of the Barn Owl, *Tyto alba erlangeri*, is also restricted to the Arabian Desert.

Several species of owls range over a wide variety of habitats including deserts. Burrowing Owls (*Speotyto cunicularia*) get a lot of attention in this book – their physiology of temperature regulation has been studied by Coulombe (1970), and they are desert inhabitants because the desert satisfies their low vegetational requirement. They are equally at home in treeless grassland, short-grass prairie, llanos, tropical savanna, sagebrush, pampas, and even alpine meadows of the western hemisphere. The situation is similar for Long- eared Owls (*Asio otus*). They are not specifically adapted to the desert but they occur commonly in arid environments because their particular habitat requirements for hunting permit exploitation of the desert. Long-eared Owls like to roost in either deciduous or coniferous trees, and may be found in the densest part of a forest, but they are specialized for hunting small mammals in open areas (Whitman 1924; Getz 1961; Marti 1976; Craig and Trost 1979). Long-eared Owls will therefore hunt mice, kangaroo rats, and other small mammals in the desert at night and then be found roosting in such places as date palms in the Israeli desert or mesquite groves in the North American desert (Hume 1991).

Elf Owls (*Micrathene whitneyi*) (Fig. 15) have so often been photographed within woodpecker holes in saguaro cacti (*Carnegiea gigantea*) that these diminutive raptors are understandably associated with deserts. Indeed, Elf Owls are most common in the Sonoran Desert of Arizona, New Mexico, and Mexico, but their habitat does extend into western pine and oak forests. An abandoned woodpecker hole, or some other hide-away for daytime roosting, is a prerequisite for these birds whose size makes protection from diurnal predators a primary consideration in habitat selection.

At least 11 other species of owls can, to variable degrees, be expected in desert habitats. Some are wide ranging species that have desert populations, others inhabit semiarid regions including or bordering deserts. Here they are mentioned in taxonomic order according to Sibley and Monroe (1990): White-faced Scops-Owl (*Otus leucotis*) of Africa south of the Sahara; Western Screech-Owl (*Otus kennicottii*) of western North America; West Peruvian Screech-Owl (*Otus roboratus*) from arid mountain regions in northwestern

Fig. 14. Hume's Owl (*Strix butleri*) is a strictly desert species of the Arabian Desert. (Illustration by Will Yandik)

Peru; Great Horned Owl (*Bubo virginianus*) of North America; Eurasian Eagle-Owl (*Bubo bubo*), a wide-ranging Palaearctic species; Rock Eagle-Owl (*Bubo bengalensis*) of Pakistan and India; Pearl-spotted Owlet (*Glaucidium perlatum*) of Africa south of the Sahara; Ferruginous Pygmy-Owl (*Glaucidium brasilianum*) of Middle America; African Barred Owlet (*Glaucidium capense*), whose acacia veld habitat borders the Namib Desert; Little Owl (*Athene noctua*) whose wide range includes the Sahara, Arabian, and Gobi Deserts; and Spotted Owlet (*Athene brama*) of southern Asia.

Fig. 15. Elf Owls (*Micrathene whitneyi*) have often been portrayed within saguaro cactus cavities where they escape daytime heat and avoid predators. (Illustration by Will Yandik)

5.1 Predatory Adaptations in Owls

With hooked bills and sharp arched claws, the raptorial adaptations of owls parallel those of hawks and falcons. The similarities between owls and hawks are so pronounced that these groups have been, at times, considered phylogenetically linked (Cracraft 1981). Molecular biologists assure us these days that the similarities between owls and hawks are, however, a case of convergence and that they represent two different branches of avian evolution (Sibley and Monroe 1990). Significant anatomical differences between them include (1) the lack of a crop in the digestive system of owls, (2) the fusion of two bones (maxillopalatine processes) in the roof of the mouth of hawks (with the exception of one genus of kites), and (3) the location of the voice box (syrinx), which in hawks is at the junction of the trachea and bronchial tubes, while owls have it divided into two parts and located down in the upper part of the bronchial tubes (Fig. 16).

Owls are recognized by their large head and flat, disk-like face with a downward pointing bill. The facial feathers are adapted to direct sounds to the large ears located behind the eyes. Soft feathers and a special comb-like

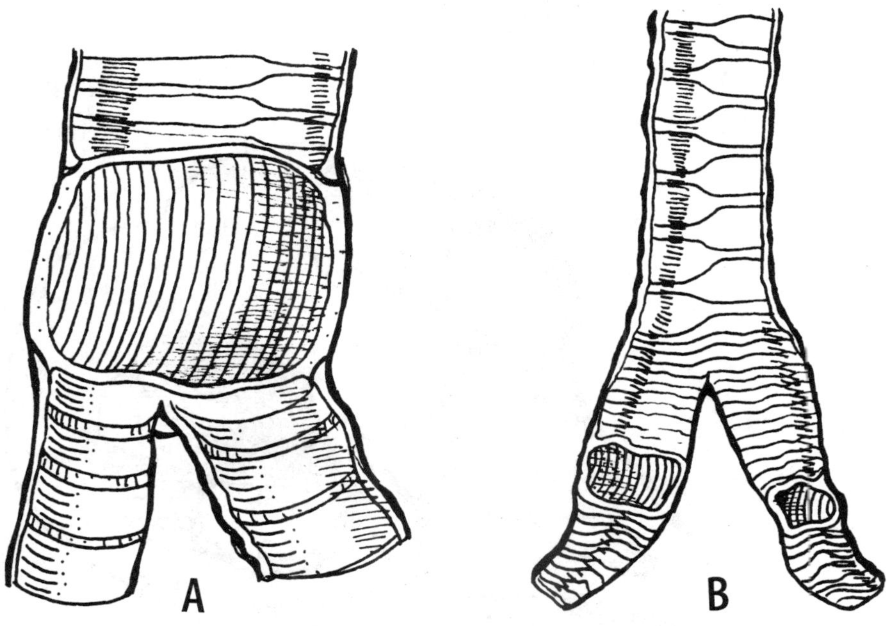

Fig. 16A, B. The syrinx of owls (B) is in two parts, located in the upper region of the bronchial tubes, while in hawks (A) it is located at the junction of the trachea and bronchial tubes. (Illustration by Will Yandik)

modification of the anterior edge of the leading primaries allow owls to fly silently while in pursuit of prey. Large eyes, each enclosed in a bony socket (sclerotic ring) with a shape reminiscent of NASA's Mercury space flight capsules, augment night vision for hunting. Owls that search for prey over long distances, such as Snowy Owls (*Nyctea scandiaca*), tend to have evolved elongated sclerotic rings (Hume 1991; Fig. 17). An examination of the sclerotic rings of diurnal desert owls might reveal them to be elongated as well, since they probably search for prey over long distances too.

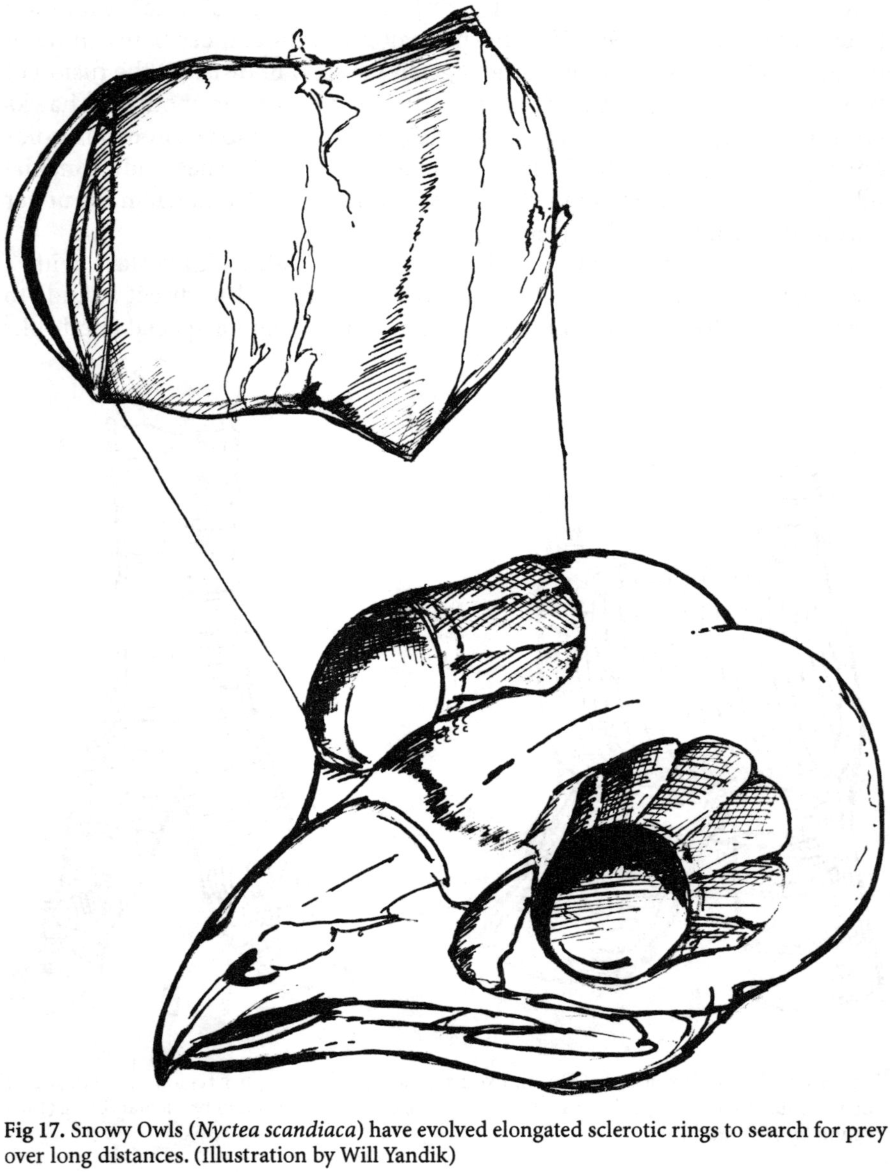

Fig 17. Snowy Owls (*Nyctea scandiaca*) have evolved elongated sclerotic rings to search for prey over long distances. (Illustration by Will Yandik)

The inability of the eyes to move within the sockets is accommodated by an astonishing ability to turn the head. In surveying their environment owls can turn their head about 135° to the right or left, swing their head back so that the crown is directed toward the tail, or tilt their head upside down to the side so that the crown faces the ground. Evolution has provided the feet of owls with reversible outer toes so that grasped prey can be held more securely. This suite of characteristics underline how effective the predatory behavior of owls has been in the evolutionary selection of physical adaptations that make owls skillful raptors.

The occupation of abandoned woodpecker holes in saguaro cacti on the Sonoran Desert may be considered an adaptation that assists predators in the procurement of prey. Several species of raptors, including Elf Owls, Western Screech-Owls, and American Kestrels, inhabit the cavities and feed upon the mice, insects and songbirds that are attracted to the saguaro nectar and fruit (Toops 1994).

5.1.1 Nocturnal Species

Of all the raptors, one would expect owls to be well represented in deserts. Indeed, in many ways they seem especially well suited – perhaps even the term "preadapted" is appropriate – to surviving the rigors of the desert environment. By hunting at night and utilizing rock crevices, and other such protective roosting areas, during the day, they can avoid midday heat and reduce evaporative water loss. Further, many desert denizens are active only at night, so the food supply for owls may even be more diverse than for diurnal predators. On warm summer nights in Arizona the nocturnal Elf Owl finds the desert crawling with the beetles, grasshoppers, and crickets that are his principal prey (Alcorn 1986). In fact, the most common bird of prey in the North American deserts is the Western Screech-Owl (Miller and Miller 1951). They greatly outnumber the diurnal birds of prey, the American Kestrels, Loggerhead Shrikes, and Greater Roadrunners, and demonstrate that the desert offers its nighttime inhabitants much more favorable conditions for life.

The nocturnal adaptations of owls allow them to take advantage of the more favorable environment of the desert night. These adaptations include a reflective layer in the retina for proficient nighttime vision and large asymmetric ears for accurate directional hearing (Burton 1973). The Barn Owl, which has representatives in deserts, and is one of the most nocturnal of all owls, has been demonstrated to be able to hunt by sound alone (Payne 1971; Knudsen 1981; Konishi 1983).

A strictly desert species, Hume's Owl, whose calls can be heard at night in February and March (Mendelssohn et al. 1975), was until recently considered the rarest owl in the world, since it was known only from three specimens (Burton 1973). According to Mendelssohn et al., Hume's Owl does not seem to be rare in the valley of the Jordan River and the Dead Sea area of Israel and Jordan, where several more specimens were taken. In the Jordan River and Dead Sea region it seems to be associated with pools or springs of water. In 1982 it was discovered in the Arabian Desert of eastern central Egypt and its known range suddenly expanded into Africa (Goodman and Sabry 1984). Owl pellets discovered in this region of Egypt may provide insight into the diet of Hume's Owl (Goodman 1986), but one can only hope that its physiology will someday be forthcoming and new insights into desert adaptations in owls will be revealed.

Quick reflexes seem to be an adaptation in many desert raptors. In order to procure a variety of prey, a Pallid Scops-Owl will dart about the deserts of Asia with nimbleness reminiscent of a roadrunner. The owl will at one moment be picking insects off foliage; in the next moment he will pounce on a beetle on the ground; in the evening he may take to the air to grab moths fluttering about; and then he may swoop to snatch bats in mid flight (Hume 1991). Along the southern edge of the Sahara Desert the Pearl-spotted Owlet also outmaneuvers bats with amazing agility. Because bats are so quick, few predators are able to catch them in flight. The few raptors that are known to be able to feed on bats are falcons and a variety of owls (Aymerich et al. 1982). Mader (1978) asserts that agility in Great Horned Owls permits them to utilize a wider range of nest sites in saguaro-palo verde flatlands in the Sonoran Desert. In addition to nesting on easy-access saguaro cacti (*Carnegiea giantea*), they can penetrate thick cover and nest in palo verde (*Cercidium* sp.) and ironwood (*Olneya tesota*) trees. In order for an owl to successfully adapt to a life on the desert, it helps to be especially spry.

5.1.2 Crepuscular Species

Crepuscular owls also avoid the heat of midday by feeding at dawn and dusk. The Ferruginous Pygmy-Owl of southwestern North America feeds mainly on small birds that are active in the early morning and late afternoon (Kaufman and Stenberg 1988). The Little Owl, including a pale desert race living in Arabia and parts of the Sahara, hunts mostly at dusk even though it is also active during the day or night (Grossman and Hamlet 1964; Hume 1991). The Spotted Owlet is a close relative of the Little Owl and seems to be primarily a crepuscular hunter. Although associated with parklands and gardens of the Indian subcontinent the Spotted Owlet is also found in desert areas of India, Pakistan, and Iran.

A notably crepuscular species, the Short-eared Owl (*Asio flammeus*) (Fig. 18), is characterized by black markings around the eyes. It has been suggested that these markings may serve to reduce the glare produced when the sun is situated low on the horizon in an open area, functioning like the black smudges that North American football players put on their cheeks during a game

Fig. 18. The Short-eared Owl (*Asio flammeus*) is characterized by black markings around the eyes which may serve to reduce glare from the sun as it hunts over open areas at dawn and dusk. (Photograph by Richard Guthrie)

(R. Guthrie pers. comm.). If so, one might expect other crepuscular species hunting on the desert to evolve this trait. The Ferruginous Pygmy-Owl, on close inspection, does indeed have a small black crescent on the medial side of the eye. But the hunting style of a Ferruginous Pygmy-Owl is distinctly different from that of a Short-eared Owl. The Short-eared Owl courses over open fields, often looking directly into the sun, whereas a Ferruginous Pygmy-Owl tends to hunt from a concealed perch (Hume 1991). The Pygmy-Owl's need for glare reduction may not, therefore, be so immediate. Little Owls tend to hunt from open perches (Burton 1973). It would be interesting to compare populations of Little Owls from deserts to populations from forests to see if there is any difference in the amount of black around the eyes.

5.1.3
Diurnal Species

Where the lack of vegetation provides little in the way of perches, some desert predators, like the Burrowing Owl, have evolved elongated legs (Fig. 19) with which they are able to stand tall and survey greater distances on a flat terrain. Of course this adaptation only makes sense to a predator hunting visually, and the Burrowing Owl is one of the few owls that is primarily diurnal

Fig. 19. The diurnal Burrowing Owl (*Speotyto cunicularia*) uses its long legs to stand tall and search for prey over flat desert terrain. (Photograph by William Cook)

(Johnsgard 1991; Haug and Didiuk 1993). The Burrowing Owl's ears are not especially well adapted to hearing at night, and its eyes lack the special adaptations of other nocturnal hunters. Still, the Burrowing Owl will avoid the intense heat of summer middays and hunt at dawn and dusk and even during moonlit nights. As a result, the Burrowing Owl is most diurnal during the cooler months (Hume 1991).

It has been pointed out that the Burrowing Owl's diet consists chiefly of diurnal species (Johnsgard 1991). This might provide an explanation as to why a bird with presumably nocturnal ancestors became diurnal even in desert situations, where it would seem to be at a disadvantage to be active during the day. The competition from diurnal raptors and the threat of predation from these same raptors are factors of considerable consequence in the Burrowing Owl's ecology and would seem to discourage diurnal activity (Johnsgard 1991). So far, however, the evidence for the conclusion that their food preferences favor diurnal activity is not convincing. In the spring Burrowing Owls prey mainly on small mammals, then switch to insects in the summer (Hume 1991). Most small mammals are more active at night. Scarab beetles, a large group of insects with nearly 1300 species in North America alone, seem to be favored by Burrowing Owls. Various species of scarab beetles may be active during the day or night. Hume (1991) points out that Burrowing Owls are also especially fond of Jerusalem crickets (*Stenopelmatus* sp.), a nocturnal insect group that spends the day under stones or in loose soil (Borror and White 1970). Owls, when given a choice, generally select prey that are easy to catch and unapt to cause injury (Marti and Hogue 1979). Within this framework some species specialize in certain prey items. Barn Owls, for example, are specialized to prey on rodents at night (Jaksić and Yáñez 1979). However, since Burrowing Owls hunt by day or night, and since desert habitats are relatively low in biomass, one would expect them to be rather indiscriminate in their food selection. This has been shown to be the case in the Atacama Desert where Burrowing Owls eat a wide range of prey items (Schlatter et al. 1980). It seems that some factor other than diet must have been at work to result in the lack of nocturnal vision and hearing in Burrowing Owls.

Fossil evidence suggests that Burrowing Owls may be among the earliest forms of owls to have evolved (De La Torre 1990). One might speculate that they represent a form that existed before nocturnal adaptations modified the rest of the owl clan. But the evidence does not seem to support this proposition either. The nearest relatives to the owls, as indicated by molecular (DNA hybridization) studies, are the goatsuckers (Caprimulgidae), which share several of the owls' adaptations for nocturnal hunting. In fact, Sibley and Monroe (1990) place all of these nocturnal groups – the frogmouths, oilbirds, potoos, nightjars, and nighthawks – in the same order as the owls. On the other hand, De La Torre suggests that the occurrence of nocturnal adaptations in owls and goatsuckers might be a case of convergence. So the questions remain: did the ancestors of Burrowing Owls lack nocturnal vision and hearing, or did Burrowing Owls evolve from nocturnal owls that lost their special adaptations for hunting at night; and if they lost their nocturnal adaptations, what selective pressures favored the switch to diurnal activity?

5.1.4 Antipredator Adaptations

Desert owls are camouflaged. A Pallid Scops-Owl looks so much like the flaking bark of a gnarled tree in a desert wadi that he is easily overlooked by predatory eyes (Hume 1991). The pale coloration of owls in desert areas can probably be best understood in terms of protection against predators as well as improving efficiency in hunting.

In deserts, where vegetation for the concealment of nestlings is minimal, owls make use of other antipredator devices for the protection of their young. The young owlets of open nest species often leave their nest long before they can fly in order to reduce the risk of detection (Whitman 1924; Burton 1973). A nest is a structure that is too easily detected by the keen eye of a raptor, and the scent of accumulated remains of food items around a nest will eventually draw the attention of a sniffing mammalian carnivore. It is best for the owlets to get out, rely on their protective coloration, and not stay too long in one place. It would certainly be advantageous for desert owlets to abandon a nest in a lonely mesquite or acacia as soon as possible. Even though most North American desert owls nest in a natural cavity or woodpecker hole in a tree or cactus, the young of these cavity nesting owls usually leave the nest several days before they can fly (Gehlbach 1986). Since owls are asynchronous hatchers, the oldest owlet of the clutch will be capable of fledging before the others, but when one leaves the nest cavity the others tend to follow, soon to be found clustered together on a nearby branch, whether they are able to fly or not.

Burrowing Owls are one of the few terrestrial birds that occupy a subterranean nest. An underground burrow is perhaps the safest nest site in a desert. These owls will use rock crevices or deserted mammalian burrows, either of which provide suitable cover in desert situations where vegetation is minimal. Although Burrowing Owls are capable of digging with their feet, they usually save work by occupying the den of a badger (*Taxidea taxus*), ground squirrel (*Citellus* sp.), or prairie dog (*Cynomys* sp.) (Alcorn 1986).

One of the most remarkable and consistent features of the nests of Burrowing Owls is that the favored nesting material is the dung of horses (Alcorn 1986). As the advent of mechanization replaced horse power with tractors, the owls adjusted to using more cow dung. I suspect that in precolonial times the Buffalo (*Bison bison*) may have been the primary source of nesting materials in North America, and the three species of *Lama* were probably the South American suppliers. An owl's burrow may be completely lined with dung throughout its length. This represents a considerable effort on the part of the bird, since the accumulation of the large quantities of dung that are found in a single burrow requires collecting over a wide area. The advantage to be gained from this tedious activity is believed to be the masking of the scent of the nest from the major nest predator, the badger. A pair of badgers are quite capable of completely digging out an owl's nest, and such predation accounts for many nest failures.

Another adaptation that seems to be a rather effective antipredator device in deserts is that, at the approach of an invader, the nestlings of Burrowing Owls emit a vocalization that sounds very much like the rattle of a rattlesnake. Since rattlesnakes can typically be found in the same sort of underground cavity as a Burrowing Owl, this technique provides an effective deterrent to many predators. Interestingly, tree cavity nestlings, such as other species of owls and certain woodpeckers like flickers, produce a similar utterance, but since a rattlesnake would not be expected to be in a tree, the same sound is interpreted as bee-like.

5.2 Temperature Regulation in Owls

5.2.1 Adaptations for Warming

Birds in general are capable of withstanding extremes of temperature that would be lethal to most humans. (Australian aborigines seem to be exceptional among humans in being able to adjust to a wide range of variation in ambient temperature without the benefit of protective clothing, but even their abilities are no match for those of birds.) It is not fully understood how tiny Golden-crowned Kinglets (*Regulus satrapa*) survive the frigid temperatures of winter in western Maine, where the thermometer often reaches -34 °C (-30 °F) but they do employ a variety of techniques including metabolic mechanisms of energy conservation, puffing out their feathers, countercurrent heat exchangers, nocturnal shelter and huddling together (Heinrich 1993). Larger birds like owls may, in contrast, appear comfortable in temperatures that are daunting to humans, as William Shakespeare implies in Love's Labour's Lost:

When icicles hang by the wall,
And Dick, the shepherd, blows his nail
And Tom bears logs into the hall,
And milk comes frozen home in pail,
When blood is nipp'd and ways be foul,
Then nightly sings the staring owl, Tu-who;
Tu-whit, tu-who - a merry note,
While greasy Joan doth keel the pot.

The common Tawny Owl (*Strix aluco*), which is considered a particularly musical owl and sings "tu-whit tu-who" all the year round but more so in autumn and winter, was surely the inspiration for the vocal element of Shakespeare's verse. The Tawny Owl is one of the many owls that begin their breeding cycle in winter in order to synchronize their production of owlets and their peak need for food with the spring emergence of an ample supply of young rodents. Many owls not only carry out their breeding activities in winter, but are particularly well adapted to cold climates. Some have exceptionally thick plumage, often making a tiny bird look much larger, as in the case of the Great Gray Owl (*Strix nebulosa*). Desert owls never have plumage as thick as that of a Great Gray Owl, but they are capable of surviving freezing temperatures. Adaptations of desert owls to cold temperatures include insulation, migration, hypothermia, and sunbathing.

5.2.1.1
Insulation

In cold deserts, like the Patagonian Desert in Argentina, the Great Basin in North America, and the Gobi Desert in Mongolia, an owl needs sufficient insulation to survive the winter. Coulombe (1970) has shown that Burrowing Owls increase their insulation and make cardiovascular adjustments to reduce heat loss in the winter. The increased weight of Burrowing Owls in winter indicates that at least part of the added insulation may be due to fat accumulation. Among the cardiovascular adjustments made by the Burrowing

Owl in winter is a reduced consumption of oxygen. This indicates that the owl's metabolic rate is reduced and fuel is being conserved. Ligon (1969) asserts that owls have lower metabolic rates than other birds to begin with, and that further, they are able to withstand drastic weight loss in order to conserve energy when food is scarce, as in the winter.

Barn Owls have less insulation than other owls and sometimes nest in burrows in cold deserts to gain the insulating value of the soil (Millsap and Millsap 1987). In northern Colorado and elsewhere in North America near the cold northern edge of their range, Barn Owls prefer to nest in cavities that they dig in an arroyo (vertical bank of an intermittent stream or "dry gulch"), rather than in a nest box, barn, or other artificial structure such as is usually considered their customary nest site. Possessing poor feather density and low fat reserves, along with a narrow thermoneutral zone, Barn Owls tend to inhabit temperate and tropical regions and migrate out of the more northerly edge of their range during the winter. Yet even during the breeding season, beginning on the ides of March in northern Colorado, the temperature can drop to -30 °C (-22 °F), making the insulating value of the soil an important factor in the survival and breeding success of ground-nesting Barn Owls.

5.2.1.2 Migration

The food supply in the desert is lower in the winter. This is especially so in cold deserts where insect prey becomes essentially nonexistent and rodents tend to hibernate out of the reach of predators. Some desert owls such as Burrowing Owls and Elf Owls are forced to migrate to warmer latitudes where insects and rodents are still active. The Elf Owl has been shown to be less able to conserve energy and have less insulation than other desert owls such as the Arizona population of the Western Screech-Owl (*Otus kennicottii cineraceus*) (Ligon 1969), so its need to migrate may be twofold involving both food supply for heat production and higher ambient temperature for heat conservation.

5.2.1.3 Hypothermia

On a circadian basis a few owls save energy by reducing metabolic heat production at night. Coulombe (1970) suggests that nocturnal hypothermia may conserve both energy and water for the Burrowing Owl, which is more diurnal in its activities than other owls. Burrowing Owls regulate their body temperature endogenously and begin to reduce their body temperature before the ambient temperature drops. The degree of hypothermia in a particular owl is dependent upon its nutritional status rather than the ambient temperature. A sufficient store of energy is probably needed to return the owl's body temperature back to its normal 38 °C (100 °F) following its period of torpor. This ability to self induce torpor has not yet been found in other owl species (Ligon 1969). The Burrowing Owl, therefore, appears to have an unusual adaptation for an owl, one that provides for additional energy conservation, but the owl must be in good health and well fed in order to take full advantage of the mechanism of hypothermia.

5.2.1.4 Sunbathing

The color of an animal can be an important factor in regulating body heat. In addition to metabolic heat production, homeotherms regulate for the conservation of heat in two ways: insulation and absorption of solar radiation (Hamilton 1973). Although there are many environmental and behavioral factors influencing the color and heat absorption of animals, black coloration has been shown to maximize solar heat absorption (Walsberg et al. 1978). Hamilton (1973) analyzes in detail the question of why desert animals tend to be black, white or pale colored and concludes that, in general, defense against predation and temperature regulation are the two most important considerations. He points out that on cool desert islands that lack predators, homeotherms are predominantly black to gain warmth from the sun; their counterparts on mainland deserts are cryptically colored to avoid predation while maintaining black skin that can be exposed for brief bouts of warming.

The difference between black birds and white birds in nonarctic environments is correlated with nocturnal behavior and their body heating strategy, according to Hamilton. Birds that seek shelter at night to keep warm do not have insulation to conserve heat and they have black plumage to absorb warmth from the sun during the day. Birds that do not seek shelter at night must retain heat with insulation and they have white plumage so as to not get overheated during the day. Most owls normally exposed to the ambient temperature during the cool night would be expected to possess insulation and not be black in color, however, diurnal desert owls, especially those that have the option to seek shelter underground at night, might be expected to have a selective pressure toward black coloration and reduced insulation.

There is some evidence that diurnal owls, in a fashion similar to roadrunners, may take advantage of solar radiation for early morning heating. Burrowing Owls face the sun in the early morning and fluff out their breast to warm themselves (Coulombe 1971). An examination of museum study skins reveals that the proximal portion of their ventral contour feathers are a dark sooty gray and may indeed be instrumental in absorbing solar radiation. According to Walsberg (1982), black skin or black plumage near the skin is effective in gaining heat for the body, while black surface plumage may not be. In a study of Phainopeplas (*Phainopepla nitens*), a sexually dimorphic passerine with black males and gray females, Walsberg determined that the heat gain by black males is minimal and could easily be ameliorated by environmental factors such as wind, or behavioral adjustments such as posture in relation to the sun. In dark-colored birds a large portion of the solar radiation received is absorbed, but this absorption is near the outer surface of the plumage so much of this heat is lost to the environment by reradiation and convection. Walsberg concludes that the Phainopepla male is black, not for thermoregulation, but for social display. Thus, for diurnal desert birds like roadrunners and Burrowing Owls who may wish to use black pigment for thermoregulation only, black skin and/or black feathering near the skin that can be exposed to the sun at appropriate times is the best solution. The outer plumage color can still be pale to provide camouflage.

5.2.2 Adaptations for Cooling

Owls that have adapted to desert life use a variety of behavioral, physical, and physiological characteristics to survive and conserve energy in a habitat where ambient temperatures easily go well above lethal levels. These adaptations involve: (1) plumage modifications to retard heat gain, (2) behavioral modifications to retard heat gain, (3) low basal metabolic rate, (4) a higher and broader thermoneutral zone, (5) hyperthermia (the ability to allow the body temperature to rise without ill effect), (6) evaporative cooling, and (7) heat radiation.

5.2.2.1 Plumage Modifications to Retard Heat Gain

Do Bedouin prefer blondes? One of the first observations that one makes concerning many desert organisms is that they are typically much lighter-colored than their humid habitat counterparts. This observation has been made so repeatedly that it has been formulated as Gloger's Law. Owls in desert situations usually conform to Gloger's Law and are pale as well. The Pallid Scops-Owl, the Pharaoh Eagle-Owl, and Hume's Owl are all described as pale colored (Grossman and Hamlet 1964; Hume 1991). Wide-ranging species of owls often have desert populations that are notably lighter-colored than other members of their own species. The desert race of the Great Horned Owl (*Bubo virginianus pallescens*) (Fig. 20) is a case in point and is even named the Pallid Horned Owl in reference to its pale plumage (Ridgway 1914). The Barn Owl is almost cosmopolitan in distribution; the palest of 34 races spanning the globe is found in the Arabian desert. The palest race of the Little Owl (*Athene noctua lilith*) lives in the same neighborhood and is found in the arid, sandy Syrian Desert east of the Mediterranean Sea (Burton 1973). On the other side of the globe, the Barred Owl has a pale desert race in Texas (*Strix varia albogilva*). Also in western North America, the Western Screech-

Owl has several desert populations recognized by their paler plumage (Ridgway 1914; Bent 1938; Miller and Miller 1951).

Pale plumage can serve two functions in deserts. It can reduce radiant heat absorption and it can defend against predation by serving as camouflage against a sun-bleached desert background. Serventy (1971) provides a detailed account of the historical development of the theoretical opinions on the topic of pale coloration in desert birds. He and Hamilton (1973) conclude that defense against predation is one of the most significant factors in determining animal coloration in the desert.

In the case of the Great Horned Owl, however, defense against predation probably is not a major consideration, since individuals of this species typically constitute the top of the food chain in their area. In fact, when a Great Horned Owl moves into a new territory, all other owls, and even diurnal raptors, may move out, causing a complete reshuffling of diurnal hunting territories in surrounding areas (Hume 1991). Such changes may not go undisputed, of course; Ferruginous Hawks and Cooper's Hawks have been observed attacking Great Horned Owls both in and out of the breeding season (Weigand 1967). Thermal regulation is the other significant reason why desert owls are paler colored; it probably has had more impact than defense against predation as a selective force in producing Pallid Great Horned Owls. Hamilton points out that reducing heat gain, as in reflecting radiant heat

Fig. 20. The desert adapted race of the Great Horned Owl (*Bubo virginianus*) of Arizona, USA is noticeably smaller and paler-colored than the forest adapted race from northeastern USA. (Photograph by William Cook)

energy, is more important than increasing heat loss in terms of "keeping cool," as it requires less expenditure of energy. Thus, temperature regulation, probably a component of Gloger's Law, as well as cryptic coloration, are believed to have conspired to result in the paling of a plethora of desert birds, including owls.

Burrowing Owls also show melanistic variation consistent with Hamilton's thermoregulation hypothesis. Juvenile owls sometimes stay outside their burrows in the midday sun and have very light plumage (Coulombe 1970). Conversely, Burrowing Owls have a dark winter plumage that absorbs more solar radiation for warmth, lowering the need for metabolic heat production and thereby saving energy.

Some desert owls have modified feathers on their toes that seem to function like pot holders. The bristly feathers allow a bird to stand on the hot desert sand while dispatching a scorpion, or some other prey item, for a midday snack. Members of the genus *Otus* are usually bare-toed, but two central American species from the arid Pacific slope, O. trichopsis (Whiskered Screech-Owl) and *O. cooperi* (Pacific Screech-Owl), have stiff bristles on their feet (Grossman and Hamlet 1964). The fully booted toes of the White-faced Scops-Owl (*O. leucotis*), a bird of the central Sahara and the dry African scrub, may also serve to protect the feet while perching on hot desert landforms.

5.2.2.2
Behavioral Modifications to Retard Heat Gain

Nocturnal activity, so characteristic of most owls, is their most significant behavioral attribute for avoiding the heat of the desert sun. However, owls that are active during the day must employ other behavioral techniques to avoid heat gain.

During late morning hours, when the temperature begins to rise above thermoneutrality, a Burrowing Owl will turn its back to the sun to shade its legs (Coulombe 1971). The long, sparsely feathered legs of the Burrowing Owl can now serve to radiate heat. At midday the owl may seek the shade of the burrow or perch high above the desert floor to avoid the heat from solar reradiation reflecting off the ground and the hot air accumulating at ground level. While on this perch, the owl extends its wings forward and outward to form a heat shield to reflect the solar radiation away from the body. Thus the owl can provide some shade for itself. While the wings are held away from the body, air can circulate more easily about the torso to facilitate cooling. Furthermore, the contour feathers of the body are compressed to reduce their insulation value. At this time, all physical activity essentially ceases and Ligon (1969) also reports that the owls usually close their eyes, although any thermoregulatory advantage of that behavior is unclear.

5.2.2.3 *Basal Metabolism*

Metabolic activity produces heat. In a hot situation it is advantageous to produce less body heat. That is one reason why desert inhabitants become inactive during the midday; inactivity reduces one's metabolic rate and internal heat production.

In addition to reducing metabolic heat by behavioral means, some desert owls have evolved physiological adaptations to reduce metabolic heat production. Great Horned Owls in the Sonoran Desert of Arizona have evolved a lower metabolic rate than the Spotted Owls (*Strix occidentalis*) that live in the cooler forested regions adjacent to the Sonoran Desert (Ganey et al. 1993). The study by Ganey et al. is a persuasive comparison, demonstrating that desert and forest habitats can have a noticeably different impact on the evolution of physiological adaptations in two owls from the same geographic region.

5.2.2.4 *Thermoneutral Zone*

An animal's thermoneutral zone is the range of ambient temperatures at which the animal does not have to expend energy for temperature regulation. In hot deserts some owls are adapted to tolerate very high daytime temperatures as well as quite low nighttime temperatures. Great Horned Owls from the Sonoran Desert of Arizona have a thermoneutral zone of 20.3-32.2 °C (68.5-90.0 °F) in contrast to Mexican Spotted Owls that inhabit forested regions of Arizona and have a thermoneutral zone of 17.0-25.2 °C (62.6-77.4 °F) (Ganey et al. 1993). The thermoneutral zone of the desert-adapted Great Horned Owl is both higher and broader than that of the Spotted Owl. This allows the Great Horned Owl to withstand the wider and higher ambient temperature range of the desert with less expenditure of energy for temperature regulation.

5.2.2.5 *Hyperthermia*

The zone of thermoneutrality for a Burrowing Owl is 25-37 °C (77-99 °F) (Coulombe 1970). When the ambient temperature goes above or below this range, oxygen consumption in the owl increases – an indication that the bird is expending energy to regulate its body temperature. As the ambient temperature goes down, oxygen consumption rises slowly, in a linear fashion (more slowly in winter owls than in summer owls) and the body temperature is maintained at 38 °C (100 °F) even in freezing weather. As the ambient temperature rises above 37 °C (99 °F), however, oxygen consumption increases very rapidly, in a parabolic curve, and the body temperature rises.

This rise in body temperature is a very important adaptation to desert heat since it reduces the need for evaporative body cooling and cuts down on water loss. Heat is radiated passively to a relatively cooler environment. Thus, hyperthermia reduces the metabolic expense of cooling the body when the ambient temperature is high. Body temperatures up to 44 °C (111 °F) have been recorded for desert-dwelling Burrowing Owls (Coulombe 1970) and Elf Owls (Ligon 1969). Great Horned Owls also exhibit hyperthermia (Chaplin et al. 1984). Hyperthermia allows a desert owl to tolerate moderately high air temperatures, but when the ambient temperature rises above a certain limit of tolerance, the body must be kept cooler than the surrounding air. Desert owls have been shown to be better than nondesert owls at keeping their body temperature below ambient temperature when the ambient temperature rises above the limit of tolerance (Ligon 1969). Desert owls have apparently evolved some physiological modifications that allow them to better tolerate desert heat.

Despite their increased tolerance of heat, desert owls, like most birds, are noticeably uncomfortable when placed in situations of thermal stress; death can occur at temperatures just a few degrees above the zone of thermoneutrality. Ligon (1969) has demonstrated that Elf Owls from the desert, like Northern Saw-whet Owls (*Aegolius acadicus*) from the temperate forest, may die within a few hours after being exposed to an ambient temperature between 41-42 °C (106-108 °F) although the Elf Owl seems able to survive a longer exposure to this heat. Other species of birds can tolerate higher temperatures without ill effect. Thus, owls, even desert inhabitants, seem to have lower lethal temperatures than other birds. It seems that, in terms of physiological adaptations for keeping cool, owls, as a group, are not particularly well suited to diurnal activity in hot deserts. For many owls behavioral adaptations such as nocturnal activity are probably a vital component in their ability to exist in deserts at all.

5.2.2.6
Evaporative Cooling

Gular flutter and panting take advantage of the cooling properties of water evaporation from the respiratory tract to reduce the body temperature. Gular flutter allows evaporative cooling to take place in the buccal cavity and requires less energy than panting. Panting moves air through the respiratory tract and therefore engages additional surface area for evaporative cooling. Gular flutter and panting have been recorded for Barn Owls (*Tyto alba*), Western Screech-Owls (*Otus kennicottii*), Eastern Screech-Owls (*Otus asio*), Whiskered Screech-Owls (*Otus trichopsis*), Great Horned Owls (*Bubo virginianus*), Elf Owls (*Micrathene whitneyi*), Burrowing Owls (*Speotyto cunicularia*), and Northern Saw-whet Owls (*Aegolius acadicus*) (Bartholomew et al. 1968; Ligon 1969; Coulombe 1970). Of these species the desert birds were able to keep their body temperature below ambient temperature, at least in low humidity which facilitated evaporative cooling, however, the nondesert owls were unable to keep their body temperature below ambient even while employing gular flutter and panting, presumably because they possess thicker insulation. In another study, desert-adapted Great Horned Owls were able to dissipate nearly 100% of their metabolic heat through respiratory evaporation, in contrast to forest-adapted Spotted Owls which were only able to dissipate 70-80% (Ganey et al. 1993).

Gular flutter is not as easy or as efficient for owls as it is for other birds (Ligon 1968). Under laboratory conditions an Elf Owl subjected to heat stress to induce gular flutter closed its eyes, panted with conspicuous movements of the abdomen, opened and closed its mouth with snapping motions, and attempted to escape from the cage. Test results indicated that this Elf Owl could only dissipate 0.88 times its metabolic heat, compared to a Common Poorwill (*Phalaenoptilus nuttallii*) which could dissipate 1.70 times its metabolic heat. The poorwill, it should be noted, has a particularly large surface area in its buccal cavity for evaporative cooling and might be expected to be rather efficient at cooling by gular flutter. Clearly though, if their metabolic heat production is greater than their ability to dissipate it by evaporative techniques, desert owls are not physiologically capable of withstanding much heat stress. Fortunately, in the wild, Elf Owls seldom need to resort to gular flutter. They simply find suitable shade to avoid being overheated. In this respect male Elf Owls do make a thermoregulatory concession to the desert. Whereas male Elf Owls inhabiting cooler mountain areas seek shade in oak or juniper thickets, males in the desert situations, with little shade available, join a female in a cavity, such as the traditional woodpecker hole in the saguaro cactus, for the duration of the daylight hours.

Elf Owls and Burrowing Owls produce a mucus secretion during gular flutter (Ligon 1968; Coulombe 1970). The purpose of this secretion was not ascertained by these authors but it seems that it might be valuable in preventing the surface of the buccal cavity from drying out during gular flutter.

5.2.2.7 Heat Radiation

The ability to radiate excess heat is maximized in small thin bodies. A higher surface-to-volume ratio gives homeotherms living in hot deserts a selective advantage. The fennec (*Fennecus zerda*), a desert fox inhabiting the Sahara and Arabian Deserts, for example, is well known for its long ears that serve as sites for heat radiation and its small slim body that does not conserve excess heat (Walker 1975). For an owl, a small size would seem to offer considerable advantage in terms of reducing heat conservation in a bird form that is otherwise adapted to conserving energy. Indeed, desert races of wide-ranging owl species are smaller. Great Horned Owls in North America (Fig. 20) and Eurasian Eagle-Owls both have "diminutive" populations in desert regions. For example the average winglength of a female Great Horned Owl of the temperate forest race is 366.3 cm, while the average winglength of females of the desert race is 362.8 cm (Ridgway 1914). The smallest race of North American screech-owls (*Otus kennicottii xantusi*) is also a desert inhabitant in southern California and Mexico (Miller and Miller 1951).

Whereas mammals tend to modify ear size for thermoregulation, birds tend to use their feet and legs to modify their body temperature. Although owls have not been observed to cool their legs the way that some vultures do, by urinating on them, some desert species are noted to have less feathered insulation on their legs and feet. The long, sparsely feathered legs of the Burrowing Owl serve to radiate heat. The unfeathered toes of Hume's Owl set this desert species apart from the closely related Tawny Owl and are also very likely an adaptation for increased heat radiation. Curiously, this adaptation is the opposite of that found in the White-faced Scops-Owl, a desert bird with feathered toes that may serve as protection against heat gain.

The compression of contour feathers on the torso reduces their insulation value by eliminating the air trapped within the feathers. This behavior facilitates heat radiation and has been observed in such desert owls as Elf Owls, Whiskered Screech-Owls, Western Screech-Owls, and Burrowing Owls (Ligon 1969; Coulombe 1971).

5.3 Water Conservation in Owls

5.3.1 Sources of Water

Desert owls, like other terrestrial birds, obtain water from three sources: drinking free water, eating water-rich food, and the production of metabolic water through cellular respiration.

5.3.1.1 Free Water

When fresh water is available, desert owls will drink it. And like other species, they will drink more in hot weather. Burrowing Owls in the field have been observed to drink more frequently at ambient temperatures above 30 °C (86 °F). In laboratory experiments Burrowing Owls drank more water at 41 °C (106 °F), a temperature that puts mild heat stress on the owl, than at 35 °C (95 °F) which is in the zone of thermoneutrality. Furthermore, at 41 °C, the owls also drank at night and showed nocturnal activity that was not present at the lower temperature (Coulombe 1971). The nocturnal drinking seems to be related to nocturnal activity which, in the case of a Burrowing Owl, a fundamentally diurnal species, occurs mostly in the summer, when high ambient temperatures impair activity during the day.

5.3.1.2
Water-Rich Food

Owls, like other raptors, have the advantage of eating meat, a food source with a very high water content which can in many cases completely satisfy a bird's water intake requirements. Bartholomew and Cade (1963) determined that many raptors can maintain, or even gain, weight on a diet of flesh without drinking water. The birds that they studied included many species of owls: Snowy Owl (*Nyctea scandiaca*), Great Horned Owl (*Bubo virginianus*), Northern Hawk Owl (*Surnia ulula*), Boreal Owl (*Aegolius funereus*), and Eastern Screech-Owl (*Otus asio*). Schildmacher (1936) reported a Tawny Owl that lived in captivity for several months without drinking water. Water makes up about 75% of the protoplasm of the cells of most animals (Berger 1961). The consumption of flesh therefore provides raptors with a substantial source of water and makes them particularly well suited to survival in deserts.

Desert owls, like Hume's Owls and Pallid Scops-Owls, are more common in riparian areas where water is available (Mendelssohn et al. 1975). However, in areas where the owls do not have direct access to free water they can survive on ground water that is indirectly available through the food chain. The White-faced Scops-Owl, a wide-ranging resident of Africa, normally avoids the most arid areas, but it has been reported from the dry Kalahari bush where it is found in thorny trees along the banks of dry watercourses (Hume 1991). Deep-rooted vegetation found near dry watercourses has access to water deep in the ground. Roots of acacia trees in Africa have been found 8 m (25 ft) below the surface and mesquite trees in North America send roots down 30 m (100 ft) (Leopold 1967). Other desert plants are capable of storing water for long periods following the infrequent floods that transform the desert into a vibrant, although ephemeral, garden. Rodents and insects feed on the succulent tissues of these drought-resistant plants. Then, when an owl consumes a mouse, gerbil, or grasshopper, the bird will be obtaining water that was, at least in part, originally salvaged by the local vegetation.

5.3.1.3
Metabolic Water

No special physiological modifications are known to exist in the metabolic production of water in desert owls. During cellular respiration desert owls, like other birds, presumably produce: (1) 1.07 g water for every g fat burned, (2) 0.56 g water for every g carbohydrates burned, and (3) 0.40 g water for every g protein burned (Gill 1990).

5.3.2 Water Conservation in the Excretory and Digestive Systems

Owls, like other birds, lose water from four regions:
(1) The excretory system, (2) the digestive system, (3) the respiratory system, and (4) the skin. The digestive and excretory systems are linked together in birds because the intestine and the ureters empty into a common cloaca. Birds lose less water than mammals from the excretory system because the kidneys and ureters absorb much more water and excrete a semisolid uric acid rather than urea as mammals do (Welty 1982). Further, the rectum, ceca, and cloaca reabsorb water from the digestive tract to further reduce water loss from the digestive and excretory systems. There is some evidence that the ability to reabsorb water in the digestive system may be better developed in desert species (Thomas 1982).

Ornithology is full of opportunities for research leading to new knowledge. Adaptations for water conservation in desert owls such as the little known Hume's Owl, for example, are a wide-open, unstudied topic that could potentially provide new insights into desert physiology.

5.3.3 Water Conservation in the Respiratory System

Gular flutter uses moisture from the mouth (buccal cavity) and panting uses moisture in the respiratory tract for evaporative cooling. In Burrowing Owls and Elf Owls it has been noted that a mucus secretion is produced in the buccal cavity during gular flutter (Ligon 1968; Coulombe 1970). Although the whole idea of gular flutter is to use water for cooling, the mucus may retard the rate at which water is evaporated while preventing the mouth from drying out, and may thereby reduce the effectiveness of gular flutter as a cooling technique in these species. At any rate, desert owls will not, if possible, resort to gular flutter and panting for cooling since these processes deplete precious water reserves. Instead they attempt to avoid the need for evaporative cooling.

The night is a cool world. Nocturnal owls, while participating in the night life in their area and partaking of its cuisine, conserve water by avoiding daytime heat and the need for evaporative cooling. Gular flutter and panting use water from the respiratory system for evaporative cooling; they appear to be emergency measures in owls and are not particularly effective in this order of birds anyway. At high humidities even a desert owl cannot keep its body temperature below ambient temperature (Ligon 1969). So a desert owl is especially wise to seek shelter during the day.

Diurnal owls also avoid the need to use water for evaporative cooling as much as possible. A Burrowing Owl will hunt during the daylight hours when ambient temperatures are below, at, or only slightly above thermoneutrality. The winter season provides a long cool period for diurnal hunting. The average January temperature for the deserts of Mexico and southwestern United States is 10-20 °C (50-68 °F) (Hammond 1992). In the summer, however, when midday temperatures may reach 49 °C (120 °F), Burrowing Owls will be active at dawn and dusk but not midday and will need to supplement their diet by working into hours of darkness even though, unlike most owls, they are not particularly well adapted to do so.

Hyperthermia is tolerated to a certain degree in owls and this saves water (Coulombe 1970). By allowing their body temperature to rise with the ambient temperature, owls do not waste water on evaporative cooling. Desert owls have been shown to tolerate hyperthermia better than other owls. The smaller bodies of desert owls and their lower levels of insulation also allow them to radiate heat better than other owls so they rely less on evaporative cooling.

5.3.4 Water Conservation in the Integument

The feathers and scales of birds retard evaporation from the skin and conserve water, however, birds in hot deserts rely on evaporation for cooling so they must allow a certain amount of evaporation to occur. Cooling that uses the least amount of water is desirable in the desert, and desert owls are more efficient at using water for evaporative cooling than nondesert owls.

First of all, desert owls are more efficient at using water for evaporative cooling because they have evolved physiological adaptations to reduce the amount of water needed to cool the body. Ganey et al. (1993) compared the maximum evaporative water loss of two species of owls from Arizona, desert-adapted Great Horned Owls and forest adapted Spotted Owls. The Great Horned Owls evaporated only 5.5 mg water per gram body weight per hour at 40 °C, while the Spotted Owls evaporated 6.2 mg water per gram body weight per hour at 42 °C. The desert-adapted Great Horned Owls were able to limit evaporative water loss through a combination of thermoregulatory devices including a lower metabolic rate and a higher thermoneutral zone.

Seeking shelter from the dehydrating power of the sun is the most conventional behavioral technique employed by owls to reduce evaporation from the skin. Owls customarily ensconce themselves in caves, crevasses, and cavities during the day. The Elf Owl's association with cactus cavities is legendary. The role of cavities in conserving energy and water for birds was detailed by Kendeigh (1961) and shown to be particularly important to Elf Owls by Ligon (1968).

The use of burrows by Burrowing Owls has also been shown to be a valuable adaptation for conserving water. Its value in conserving water may be even more important than its value in temperature regulation. According to Coulombe (1971), burrows may not be much cooler than the surface but they are considerably more humid. The combination of moisture from the soil and from the bodies of the owls can keep the water concentration of the air in a burrow near the saturation point. The high humidity level in the deepest part of a burrow retards evaporation of water from the owl's skin. Young chicks, not yet fully feathered, would especially benefit from this humid subterranean habitat. As long as the temperature of the air in the burrow does not rise above the range of thermoneutrality (Coulombe's measurements indicate that it probably never does), and the owl does not need to evaporate water for cooling, the burrow can serve to greatly reduce water loss from the skin.

Two Old World species that are closely related to Burrowing Owls (sometimes the Burrowing Owl is placed in the same genus with them, *Athene*) also tend to be fossorial. The Little Owl (A. *noctua*) and the Spotted Owlet (A. *brama*) appear to be short-legged versions of the Burrowing Owl, and the desert populations of these species often nest in underground burrows (Burton 1973). There is a conspicuous tendency of desert owls to use underground burrows. A burrow can provide defense against predators in a land with little vegetative cover, moderate temperatures where the ambient temperature can be lethal, and offer a humid environment where water conservation is essential to survival.

5.3.5
Water Conservation in the Breeding Cycle

The unpredictability of the desert makes it likely that in any particular year there may not be enough food and water to support all of the hatchlings in a nest. When food and water supplies are insufficient it is desirable to sacrifice some but not all of the nestlings. To assure that the nestlings perish in a sequence rather than all together, the eggs are laid and hatched at intervals so that there is a graded series of sizes among the nestlings. When food and water supplies run short the smallest nestling will be unable to obtain necessary supplies and will succumb while the larger, older siblings will live. Such asynchronous hatching is common among raptors and is recorded for White-faced Scops-Owls in the Kalahari (Hume 1991) and in Barn Owls (Soucy 1979).

CHAPTER 6

Shrikes

Shrikes, the passerine raptors, the savage songbirds, the butcher birds – these are predatory birds evolved from a thrush-like ancestor, presumably in Africa, where all but a few species exist today. Some have spread into Europe and Asia and two have found their way to the New World. One, the Loggerhead Shrike (*Lanius ludovicianus*), has spread as far as Mexico and occurs in a variety of habitats including the deserts of southwestern North America. Similarly, the wide ranging Red-backed Shrike (*L. collurio*) and the Rufous-tailed Shrike (*L. isabellinus*) of the Palaearctic can be found in deserts of Asia. The only shrike that commonly occurs in both hemispheres is the Northern or Great Grey Shrike (*L. excubitor*) and it occurs in open areas from tundra to desert (Fig. 21).

Several of the African shrikes occur in deserts and arid savannas bordering deserts. They include, among many possible others, the Grey-backed Fiscal (*L. excubitoroides*), the Teita Fiscal (*L. dorsalis*), the Somali Fiscal (*L. somalius*), the Common Fiscal (*L. collaris*), the Yellow-billed Shrike (*Corvinella corvina*), the Magpie Shrike (*C. melanoleuca*), the White-rumped Shrike (*Eurocephalus rueppellii*), the White-crowned Shrike (*E. anguitimens*), the Crimson-breasted Gonolek (*Laniarius atroccineus*), the Rosy-patched Bushshrike (*Rhodophoneus cruentus*), and the Bokmakierie Bushshrike (*Telophorus zeylonus*). And new desert shrikes, such as the Bulo Burti Boubou (*Laniarius liberatus*) (Fig. 22), are still being discovered in Africa (Smith et al. 1991).

In the winter, migrant shrikes from Europe and Asia, such as the Masked Shrike (*Lanius nubicus*), the Woodchat Shrike (*L. senator*), the Red-backed Shrike, the Rufous-tailed Shrike, and the Brown Shrike (*L. cristatus*) also occur in Africa. Although they can be found in the Sahara, most seem to concentrate in the Sahel and Soudan savanna, the semiarid zones between the extreme desert and the more richly vegetated Guinea savanna near the equator (Bates 1933, Hogg et al. 1984) while others continue on to arid regions of southern Africa (Harris 1988).

Fig. 21. Shrikes are songbirds (Order Passeriformes) that have evolved hawk-like feeding behaviors. The Northern, or Great Grey Shrike (*Lanius excubitor*), shown here, hunts open areas from tundra to desert in North America, Eurasia, and Africa. (Photograph by Richard Guthrie)

The taxonomy of shrikes has become a bit confused as Sibley and Monroe (1990) have moved more than half of them from the Family Laniidae into a subfamily, the Malaconotinae, of a now expanded family, the Corvidae (crows and jays), along with Wattle-eyes and Batis, which were formerly considered Old World Flycatchers in the Family Muscicapidae.

Fig. 22. The Bulo Burti Boubou (*Laniarius liberatus*) is a newly discovered species of desert shrike from Somalia. (Illustration by Will Yandik)

6.1 Predatory Behavior in Shrikes

6.1.1 Hunting Strategy

Shrikes are unique among perching birds (Order Passeriformes). Their diet includes small vertebrates that are hunted and dispatched with falcon-like efficiency; they possess a truly raptorial feeding behavior. Shrikes typically hunt from an open perch, but they will also flush prey by flying through shrubs near the ground, or hover in the air over open fields searching for any animate object smaller than themselves (Douglas 1990). A shrike may use its feet to seize small birds in flight, forcing them to the ground. The shrike will then use its hooked and notched beak to sever the base of its victim's skull, precisely as a falcon would. When it is hunting mammals, a shrike will land next to its victim and use that powerful beak to strike at the base of the mammal's skull. Prey thus slain may be eaten on the spot, or carried to a larder. A larder is a thorn, or barb of a barbed wire fence, on which the victim is impaled. A Loggerhead Shrike (Fig. 23) can carry lightweight prey (19-58% of

Fig. 23. Loggerhead Shrikes (*Lanius ludovicianus*) carry heavy prey items with their feet but must use their bill to tear flesh. (Photograph by Tony Leukering)

the shrike's body weight) in its beak but heavier prey (61-131% of the shrike's body weight) are carried with the feet (Yosef 1993).

The impaling of prey is a unique behavior, and has given rise to the nickname of "butcher bird" for shrikes, particularly of the genus *Lanius*; this behavior seems to play several roles in the bird's life history. Impaling assists individuals in the process of feeding, it attracts mates, and it conserves a female's energy for nesting activities.

Shrikes may have the beak of a falcon but they did not evolve talons; the feet of a shrike, although fairly strong (shrikes can carry victims heavier than themselves in their feet), are too weak to tear flesh. A thorn is, therefore, used as a handy device to hold their meaty food immobile, while they tear off portions with their beak.

The collection of impaled food items made by a male is attractive to prospective mates (Fig. 24), and helps determine mate selection (by the female) and reproductive success (Yosef and Pinshaw 1989). Male Northern Shrikes

Fig. 24. The collection of food items, made by male shrikes, helps determine mate selection and reproductive success. (Illustration by Will Yandik)

increase the size of their cache of impaled food items just prior to the breeding season. A male's cache reaches its peak when a female has settled in his territory and breeding has commenced, that is, when a nest has been completed and eggs have been laid. Males with larger caches mate first and sire more offspring. The size of the food cache then declines quickly as females and young are fed from the stores in the larder.

Females are attracted to males with large food caches. Factors that have lead to the selection of this behavior probably include the likelihood that large caches indicate abundant resources in that male's territory, and that the male is a "good provider," that is, he is proficient at finding food. Yet, there may be more involved in the need to create a food cache. During the breeding season the need for food energy increases. Incubation responsibilities may prevent the female from having enough time and energy to hunt for food for herself, so food caching could be a device to conserve the female's energy for nesting duties (Applegate 1977). The male is required to obtain food for the whole family, and as a result of these additional demands on the male as the food hunter, he is released from any incubation duty. Still, in order to have enough food to make it through the breeding season, he must also build up a larder of stored food prior to the breeding season.

Desert shrikes may differ from shrikes of other habitats in their method of acquiring the ability to recognize suitable impaling devices (Smith 1972). Nondesert shrikes, in a habitat with predictable holding devices, are hatched with an innate knowledge of what that device is, while desert shrikes, in a more variable habitat, must learn, after hatching, what the most available holding device is and how to use it. Red-backed Shrikes and Woodchat Shrikes of Eurasia inhabit areas with many thorny plants such as hawthorn (*Crataegus* spp.). The young are hatched with an innate ability to impale prey with a wiping motion directed toward a thorn. Northern Shrikes living in a thornless habitat produce young with an innate ability to use forks of branches to hold prey. In contrast, the Loggerhead Shrike of the American deserts inhabits a more variable habitat, and the young have to learn which device is most available to hold prey: thorn bushes, forked branches, barbed wire, or other suitable device (Fig. 25). This learning takes place during a special period in the young bird's development, between the 20th and 30th day of life.

Learning how to capture, kill, and impale prey is essential to the survival of a young shrike. Studies have shown that chemical poisons applied to the environment can impair this learning process. Loggerhead Shrikes contaminated with dieldrin take longer to learn how to kill mice (Busbee 1977).

6.1.2 Euryphagy

Like other desert "raptors", Loggerhead Shrikes take a very wide range of prey items, including carrion in the winter when insect prey is less available.

Fig. 25. Loggerhead Shrikes (*Lanius ludovicianus*) of the American deserts learn to impale prey on thorns or use forked branches, depending upon which is available, at about the fourth week of life. (Illustration by Will Yandik)

Shrikes may even eat a wider selection of prey items than hawks; as passerines they need more food per gram of body weight than large raptors (Kendeigh 1970). Shrikes also process their food rapidly, digesting an entire mouse in 3 hours (Gill 1990).

Studies, including examinations of stomach contents, have revealed that Loggerhead Shrikes eat: mice, kangaroo rats, ground squirrels, small birds, American Coot (carrion?), flat-tailed horned lizards, round-tailed horned lizards, desert horned lizards, bleached horned lizards, greater earless lizards, prairie lizards, snails, scorpions, spiders, bees, wasps, ants, horseflies, deerflies, crane flies, maggots, cutworms, darkling beetles, ground beetles, dermestid beetles, blister beetles, click beetles, longhorn beetles, scarab beetles, cicadas, lubber locusts, bugs, dragonflies, ant lions, sand crickets, field crickets, and grasshoppers (Knowlton and Harmston 1944; Bent 1950; Bundy et al. 1955; Pianka and Parker 1975; Reid and Fulbright 1981; Turner and Medica 1982). In desert habitats where prey items are scarce, a predator must eat whatever can be found.

6.1.3 Prey Selection

The study of prey selection still needs much work. Some authors suggest that predators, when given a choice, prefer large items; others say small items are preferred; and others say that conspicuousness, activity levels, and search images are perhaps more important factors (Kaufman 1973; Mueller 1974; Hils 1985).

Slack (1975) suggests that shrikes tend to select prey that are small enough to be easily caught, thus requiring less energy expenditure. A good rationale for the small prey approach is recounted in Bent (1950), where a tale is told of a Loggerhead Shrike who, attempting to subdue a garter snake, became entangled and nearly strangled by the snake. The potential reward of a large meal suddenly became a life-threatening force. Still, there is an advantage to capturing a large prey item. It would provide food for a much longer period of time (Miller 1931).

In another study of shrikes, birds preferred inconspicuous prey (native colored mice over laboratory white mice), presumably because the native mice fit the search image of the bird (Kaufman 1973).

6.1.4 Territorial Behavior

The rustling sounds of a colony of mice, a thorny tangle of shrubbery, and a sun-bleached snag on a nearly dead acacia or mesquite makes a shrike feel right at home. A good shrike territory will provide food, cover, and lookout points (Durango 1956). Food and cover are necessary for survival and reproduction; lookout points are needed for territorial defense and as hunting perches. Nonmigratory shrikes tend to be strongly territorial and will defend good territories year-round (Marshall and Cooper 1969). Territorial boundaries of the Grey-backed Fiscal in Uganda are defended by both sexes and are marked by prominent perching and display stations (Banage 1969). Threat displays by shrikes have been described as dancing behavior, and indeed, the repelling of rival conspecifics is a part of the pair bonding ritual leading to successful mating.

These home territories are vital to the shrikes – all foraging, roosting, and breeding activities take place within them. The birds will not leave their territory unless the food supply runs out (Durango 1956). Occasionally, a male may attempt to annex new territory by fighting off neighbors, and thereby acquire additional food resources without moving to an entirely new territory.

Short grass makes hunting easier. Shrikes have to exert greater effort in the form of hovering and chasing prey in high vegetation. Loggerhead Shrikes in Florida adjust by moving to areas of low vegetation within their territories on an annual cycle to increase their food-gathering efficiency (Yosef and Grubb 1993). So, a prime territory for a shrike would have enough habitat diversity to offer an area of low vegetation throughout the year.

Male and female shrikes of the same species are normally indistinguishable by plumage. Curiously, however, some African shrikes have evolved sexual dimorphism. The adaptive value of this phenomenon is not clear, but it may function somehow in territorial defense (Grimes 1979). In the Yellow-billed Shrike, flank patches differ in color between males and females. These patches, normally concealed, are exposed during territorial disputes and displays at the nest.

6.2 Breeding Behavior in Shrikes

6.2.1 Cooperative Breeding

Cooperative breeding behaviors occur in birds from many habitats and ecological niches (Skutch 1961), and they may arise as a result of social rather than ecological factors (Dow 1980), but assistance in raising nestlings may have particular advantages for desert predators (Mader 1979). Food resources for desert predators, such as shrikes, are scattered and unpredictable, so more than one hunting adult may be vital to finding enough food items to feed a family. Furthermore, vegetative cover for nest sites tends to be minimal in deserts, so one parent may be required to attend a nest most of the time, to defend against other predators.

Several species of African shrikes found in arid areas live in groups throughout the year, and are known, or suspected, to be cooperative breeders: the Yellow-billed Shrike, the Magpie Shrike, the Grey-backed Fiscal, the Long-tailed Fiscal (*Lanius cabanisi*), the White-rumped Shrike, and the White-crowned Shrike.

The Yellow-billed Shrike, an inhabitant of the arid savannas south of the Sahara, seems to require nest helpers for successful breeding. A study in Ghana revealed that breeding units had an average of 12 members and ranged from 5 to 25 members (Grimes 1980). Groups of less than five members did not breed, indicating that helpers were a requirement for successful reproduction. Nonbreeding members of the group assist in constructing the nest, bring food to the incubating female, and feed the young when the breeding female is away from the nest.

The Magpie Shrike is a close relative of the Yellow-billed Shrike and inhabits acacia savanna bordering the Namib and Kalahari Deserts. It too is a cooperative breeder but to a more limited extent. Groups of as few as three individuals nest successfully, with two males attending a female (Nichol 1964). Although groups average around 6 members and may include as many as 11, only one female incubates the eggs and not all of the group members assist the male in providing food for the female (Harris 1988).

The White-rumped Shrike inhabits dry, open areas of eastern Africa. Two or more females have been observed taking turns incubating eggs (Moreau and Moreau 1939; Skutch 1961). The closely related White-crowned Shrike is found on the Namib and Kalahari Deserts and surrounding savannas where observations of three to five adults attending single nests attest that members of this species, at least occasionally, breed in groups (Winterbottom 1971; Grimes 1976; Harris 1988)

Cooperative breeding behavior among shrikes may have originated among populations in which young birds were retained on their natal territory while parents raised new broods. This phenomenon may be a preliminary step on the path to cooperative breeding, and has been observed among Common Fiscals in Ghana (MacDonald 1980). Young birds stay with their parents a long time – up to 7 months. This extended postnatal lingering at home may result in having young from as many as three clutches still on their parents' territory. The young Common Fiscals have not been observed to assist their parents or siblings in any way, yet the retention of young on parental territories may have been an important precursor in the development of cooperative breeding. Probably as a result of keeping these close family ties, two female Common Fiscals sometimes lay their eggs in the same nest (Marshall and Cooper 1969). Two females using the same nest may also be a step in the evolution of the cooperative breeding behaviors that have developed in other species of African shrikes.

Grey-backed Fiscals are a species that has adopted a cooperative breeding strategy – in Kenya they have greater reproductive success when they breed in large groups (Zack 1986). They occupy a habitat that is small in size, yet offers better food and cover throughout the year than the areas occupied by the Common Fiscal (Zack and Ligon 1985). Habitats that are restricted in size, yet stable (stable means providing food and cover throughout the year) may be another factor that favors cooperative breeding behaviors. Young birds may stay on their parents' territory because it offers the only source of abundant food in the region. It has been suggested that the evolution of cooperative breeding may be encouraged by habitat saturation, as would occur when an expanded family group stays on a territory (Selander 1964). When there are no new territories for fledglings to colonize, it may be to their advantage to stay with their parents, help raise their siblings, and hope to replace one of their parents as a breeding individual on their home territory.

If cooperative breeding really is an advantage to desert shrikes, one is left with an interesting contradiction to explain. In the Negev Desert of Israel, Northern Shrikes exhibit polygyny (Yosef and Pinshaw 1988b). A single male siring young from two females reduces the number of nest helpers for each female to less than one. Perhaps such a strategy only works if the male has a superior ability to cache food for the females and young.

6.2.2 Effect of Rain on Breeding

Rainfall in deserts heralds the growth of fresh vegetation and, subsequently, increased insect populations – such rainfall often stimulates the initiation of breeding in arid-adapted shrikes as well. In Ghana, for example, Yellow-billed Shrikes lay most of their egg clutches in February, March, and April, at the beginning of the rainy season (Grimes 1980). A bird may even anticipate the coming rains and build a nest in a leafless tree, expecting that the tree will produce leaves following the rain. If the tree fails to produce leaves, the shrike will abandon the exposed nest.

The Common Fiscal is a widespread shrike species with different races inhabiting different regions. The race in Ghana seems to be confined to wetter areas and breeds during both wet and dry (relatively speaking) months (MacDonald 1980). In more arid South Africa, however, the resident race of the Common Fiscal breeds only during the rainy season extending from August to December (Cooper 1970a,b). MacDonald (1980) contends that this is when South African birds would be expected to breed anyway, so it does not represent a behavioral modification, and that rainfall is relatively unimportant in determining the breeding season of Common Fiscals there. However, studies in Kenya have shown that rainfall does affect the physiology of Common Fiscals, and they will not breed during extreme droughts (Dittami and Knauer 1986; Zack 1986). Although the gonads of Common Fiscals will increase in size normally during severe droughts, they will not secrete the gonadal steroid hormones necessary to initiate reproductive activities.

6.2.3 Nest Building

Materials for nest construction may be scarce in arid habitats with minimal vegetation. Although most passerines tend to be rather rigid in their nest construction behavior, desert birds may have to use a wide variety of materials – anything they can find, actually – to fabricate a nest. Some of the many items used by Common Fiscals in Rhodesia for nest construction are twigs, string, grass, lint, woven cloth, small flower heads, paper, cigarette filters, rootlets, feathers, rags, moss, and insect cocoons (Marshall and Cooper 1969). Loggerhead Shrikes in Idaho are considered well adapted to deserts because they have sufficient behavioral plasticity in nest building to permit nesting on a variety of substrates (Woods 1993a). They adjust their nest construction to fit dense greasewood shrubs or open sage brush and bitterbrush, depending upon which is available. Diversity in nesting materials and construction techniques is among the fascinating behavioral adaptations that some shrikes have made to desert life.

6.2.4 Early Departure from Nest

In deserts, where cover is minimal and nests are particularly vulnerable to predation, passerines may lure young from the nest earlier than they would in other situations. Loggerhead Shrikes in the cold desert of Idaho sometimes induce nestlings to fledge early to avoid the loss of whole broods to nest predators (Woods 1993b). Northern Shrikes in the central Negev Desert move nestlings to new locations following a disturbance (Yosef and Pinshaw 1988a). Deserts provide very little cover and concealment for nests. In order to reduce the risk of predation and increase the chance of fledgling survival, some desert shrikes have adapted by removing nestlings from their nests at a very early age.

6.3 Temperature Regulation in Shrikes

The body temperature of birds is a product of the oxidation, or "burning," of fuel in the form of food – carbohydrates, fats, and proteins. Shrikes, along with other small birds, have a higher rate of metabolism than large birds, compensating for their greater surface area for radiating heat and their inability to carry as much plumage for insulation (Kendeigh 1970). As a result of their higher metabolic rate, small birds have a higher body temperature and are less likely to be stressed by desert heat. Some small birds can have a body temperature as high as 44.6 °C (112.3 °F) (Terres 1980), but the normal body temperature of most birds at rest is about 40 °C (104 °F) (Dawson and Bartholomew 1968).

A higher body temperature adapts a bird to desert heat in two ways: (1) the higher body temperature raises the bird's zone of thermoneutrality so that as the ambient temperature rises, the bird does not have to resort to cooling techniques as soon as animals with lower body temperatures. The zone of thermoneutrality for Loggerhead Shrikes is from 24.3 °C (75.7 °F) to at least 36 °C (97 °F) (Weathers et al. 1984). (2) A high body temperature facilitates heat loss, as heat is conducted to the cooler air that is in contact with the bird's body (Withers and Williams 1990).

When ambient temperatures go beyond the zone of thermoneutrality a bird must use physiological and behavioral techniques to maintain body temperature within acceptable limits. The temperature control center is located in the hypothalamus of the bird's brain (Terres 1980). In response to temperature receptors in the bird's skin, the hypothalamus will cause physiological warming or cooling activities to occur. Control of these responses occurs either directly through the nervous system or indirectly through the endocrine system. A shrike can warm itself in cold weather by fluffing up its feathers and shivering. In hot weather, on the other hand, panting allows increased cooling by evaporative water loss in the respiratory system.

Behavioral adjustments are also important for thermoregulation among shrikes. Greater activity levels increase metabolic activity and heat production. Weathers et al. (1984) have calculated the amount of energy a Loggerhead Shrike expends per day (106 kJ) and the amount of energy used for various activities (e.g., 3.87 kJ/h for preening and 23.7 kJ/h for flying compared to 1.79 kJ/h basal metabolism). Since increased activity increases heat production, desert shrikes will adjust their daily activity according to ambient temperature. During the cooler winter months, Loggerhead Shrikes will feed during the mid afternoon, but during the summer they feed in the morning and remain inactive later to avoid the intense heat of the mid afternoon (Craig 1978).

6.4 Water Conservation in Shrikes

Like that of other carnivorous birds, a shrike's diet, including fleshy vertebrate tissue, provides a substantial quantity of water (Bartholomew and Cade 1963). As long as the diet contains sufficient moisture, a shrike such as the Common Fiscal does not need to drink (Cooper 1970b). Even though Common Fiscals, described as facultative drinkers, will drink when water is available, the absence of water for long periods does not appear to cause stress. The territories of Common Fiscals are often devoid of free water for much or all of the year, yet they seldom or never leave their territories. Furthermore,

Irwin (1956) noted that shrikes never visited the desert waterholes that he studied in Bechuanaland (now Botswana), a region in South Africa that includes the Kalahari Desert.

The ability to survive in desert areas without drinking seems to be a remarkable feat, yet shrikes do not demonstrate any special adaptations for retaining water compared to other birds (Bartholomew and Dawson 1953). In fact, as small birds, shrikes lose water at a greater rate than large raptors. Shrikes also lack the nasal salt secreting gland which assists in the physiology of water absorption from the digestive system in many nonpasserines (Gill 1990). Their ability to survive well without drinking free water demonstrates that a carnivorous diet is a very effective source of water for predatory birds.

Even though shrikes do not need to drink free water, some species, such as the Common Fiscal and the Grey-backed Fiscal, are dependent upon seasonal rains to produce the vegetation and expanded insect populations needed for successful reproduction (Cooper 1970a; Zack 1986). In fact, if conditions are drier than normal, parent Common Fiscals will expel their fledglings from their territory earlier than usual so that when they (the parents) try to produce a second brood, they can use the minimal resources to feed the new nestlings (MacDonald 1980). If drought conditions become severe, Common Fiscals will not breed at all, a response that is hormonally controlled (Dittami and Knauer 1986).

Migratory birds, such as Masked Shrikes, Woodchat Shrikes, Red-backed Shrikes, Rufous-tailed Shrikes, and Brown Shrikes, many of which migrate over the Sahara, have a special need to conserve water because, in order to reduce weight before their migratory flight, they dehydrate their body tissues (Torre-Bueno 1978). Such migrants supply their water needs with metabolic water and prevent evaporative water loss by ascending to cold air where convection can carry off their body heat. They can meet their water needs with metabolic water only as long as the ambient temperature is at less than a critical level, about 7 °C (45 °F), so that there is no need for evaporative cooling.

Acknowledgements

I wish to thank the many friends and colleagues who reviewed the manuscript and contributed to its improved quality:

I owe special thanks to Susan Hajjar for her encouragement and guidance during the initial stages and planning of the book; to Jim Coe for his bird portraits; to Will Yandik for his illustrations; and especially to John Cloudsley-Thompson, my editor, for his consistant assistance, patience, and advice.

The manuscript was read by Dr. Kenneth P. Able. He provided valuable input that helped me clarify portions of the text. However, I take full responsibility for any errors or misrepresentations of the scientific literature that may appear in the text.

I thank Kay Cook, Wendell Cook, and Brenda Gerry for proofreading the manuscript.

Jim Coe drew Figs. 3 and 13. Will Yandik drew Figs. 2, 8, 9, 11, 12, 14-17, 22, 24, and 25. Photographs and diagrams were provided by John Cloudsley-Thompson (Figs. 1, 4, and 5); William E. Cook (Figs. 19 and 20); Richard P. Guthrie (Figs. 18 and 21); Tony Leukering (Figs. 6 and 23); Don Roberson (Fig. 10); and Mark Vrtiska (Fig. 7).

References

Ahrens C (1991) The good, the bad, and the ugly. Birder's World 5(2):20-23

Alcorn GD (1986) Owls. Prentice Hall, New York

Allen GT (1987) Prairie Falcon aerie site characteristics and aerie use in North Dakota. Condor 89:187-190

Amanova M (1975) Adaptive features of the relief structure of the intestinal mucous membrane in desert birds. Dokl Akad Nauk SSSR 225:508-510

American Ornithologists' Union (1983) Check-list of North American birds, 6th edn. American Ornithologists' Union, Washington DC

Applegate RD (1977) Possible ecological role of food caches of Loggerhead Shrike. Auk 94:391-392

Arad Z, Bernstein MH (1988) Temperature regulation in Turkey Vultures. Condor 90:913-919

Arad Z, Midtgård U, Bernstein MH (1989) Thermoregulation in Turkey Vultures: vascular anatomy, arteriovenous heat exchange, and behavior. Condor 91:505-514

Arnold LW (1942) The aerial capture of a White-throated Swift by a pair of falcons. Condor 44:280

Attwood R (1982) Birds sheltering in the burrows of mammals during a heatwave. S Aust Ornithol 29:7-9

Aymerich M, García de Castro RF (1982) Predación de Falco peregrinus y Falco subbuteo sobre quirópteros. Doñana Acta Vertebr 9:396-397

Baker MA (1993) A wonderful safety net for mammals. Nat Hist 102(8):63-64

Bakker RT (1986) The dinosaur heresies. Morrow, New York

Ballam JM (1984) The use of soaring by the Red-tailed Hawk (Buteo jamaicensis). Auk 101:519-524

Banage WB (1969) Territorial behavior and population in the Grey-backed Fiscal Shrike. Uganda J 33:201-208

Bang BG, Cobb S (1968) The size of the olfactory bulb in 108 species of birds. Auk 85:55-61

Bartholomew GA, Cade TJ (1957) The body temperature of the American Kestrel Falco sparverius. Wilson Bull 69:149-154

Bartholomew GA, Cade TJ (1963) The water economy of land birds. Auk 80:504-539

Bartholomew GA, Dawson W (1953) Respiratory water loss in some birds of the southwestern United States. Physiol Zool 26:162-166

Bartholomew GA, Lasiewski RC, Crawford EC Jr (1968) Patterns of panting and gular flutter in cormorants, pelicans, owls and doves. Condor 70:31-35

Barton NWH, Houston DC (1993) A comparison of digestive efficiency in birds of prey. Ibis 135:363-371

Bates GL (1933) Birds of the southern Sahara and adjoining countries in French West Africa, part 1. Ibis 75:752-780

Beal KG, Gillam LD (1979) On the function of prey beating by roadrunners. Condor 81(1):85-87

Beck BB, Engen CW, Gelfand PW (1973) Behavior and activity cycles of Gambel's Quail and raptorial birds at a Sonoran Desert waterhole. Condor 75:466-470

Bednarz JC (1987) Successive nesting and autumnal breeding in Harris' Hawks. Auk 104:85-96

Bednarz JC (1988a) A comparative study of the breeding ecology of Harris' and Swainson's Hawks in southeastern New Mexico. Condor 90:311-323

Bednarz JC (1988b) Cooperative hunting in Harris' Hawks (Parabuteo uncinctus). Science 239:1525-1527

Beiser A (1971) The earth. Time-Life Books, New York

Bent AC (1937) Life histories of North American birds of prey, Part 1. US Natl Mus Bull 167

Bent AC (1938) Life histories of North American birds of prey, part 2. US Natl Mus Bull 170

Bent AC (1940) Life histories of North American cuckoos, goatsuckers, hummingbirds, and their allies, part I. US Natl Mus Bull 176

Bent AC (1950) Life histories of North American wagtails, shrikes, vireos, and their allies. US Natl Mus Bull 197

Berger AJ (1961) Bird study. John Wiley, New York

Booth BDM (1961) Breeding of the Sooty Falcon in the Libyan desert. Ibis 103:129-130

Borror DJ, White RE (1970) A Field Guide to the insects of America north of Mexico. Houghton Mifflin, Boston

Boshoff AF, Vernon CJ (1980) The past and present distribution and status of the Cape Vulture in the Cape Province. Ostrich 51:230-250

Brook M, Birkhead T (eds) (1991) The Cambridge encyclopedia of ornithology. Cambridge University Press, Cambridge

Brower LP, Ryerson WN, Coppinger LL, Glazier SC (1968) Ecological chemistry and the palatability spectrum. Science 161:1349-1350

Brown LH, Urban EK, Newman K (1982) The birds of Africa, vol 1. Academic Press, London

Bundy RE, Meyer D, Ness J (1955) Observations on two species of lizards in the Chihuahuan Desert. Copeia 1955:312

Burton JA (1973) Owls of the world. Dutton, New York

Busbee EL (1977) The effects of dieldrin on the behavior of young Loggerhead Shrikes. Auk 94:28-35

Buxton PA (1923) Animal life in deserts. Edward Arnold, London

Cade TJ (1965) Relations between raptors and columbiform birds at a desert water hole. Wilson Bull 77:340-345

Cade TJ, Greenwald L (1966) Nasal salt secretion in falconiform birds. Condor 68:338-350

Calder WA (1968a) The diurnal activity of the roadrunner Geococcyx californianus. Condor 70:84-85

Calder WA (1968b) Nest sanitation: a possible factor in the water economy of the roadrunner. Condor 70:279

Calder WA Jr, Bentley PJ (1967) Urine concentration of two carnivorous birds, the White Pelican and the roadrunner. Comp Biochem Physiol 22:607-609

Calder WA Jr, Schmidt-Nielsen K (1966) Evaporative cooling and respiratory alkalosis in the pigeon. Proc Natl Acad Sci USA 55:750-756

Calder WA, Schmidt-Nielsen K (1967) Temperature regulation and evaporation in the pigeon and the roadrunner. Am J Physiol 213:883-889

Callopy MW (1983) Foraging behavior and success of Golden Eagles. Auk 100:747-749

Cameron AC (1938) Birds drinking in the dry interior. Emu 38:336-337

Cash KJ (1989) Three adult Swainson's Hawks tending a nest. Condor 91:727-728

Chaffie N (1994) Flying detox plants. Conservationist 48(3):10-13

Chaplin SB, Diesel DA, Kasparie JA (1984) Body temperature regulation in Red-tailed Hawks and Great Horned Owls: responses to air temperature and food deprivation. Condor 86:175-181

Chen DM, Collins JS, Goldsmith TH (1984) The ultraviolet receptor of bird retinas. Science 225:337-340

Chew RM (1961) Water metabolism of desert-inhabiting vertebrates. Biol Rev 36:1-31

Clark RG, Ohmart RD (1985) Spread-winged posture of Turkey Vultures: single or multiple function? Condor 87:350-355

Cloudsley-Thompson JL (1977) The desert. Putnam, New York

Cloudsley-Thompson JL, Chadwick MJ (1964) Life in Deserts. Foulis, London

Cobb S (1960) A note on the size of the avian olfactory bulb. Epilepsia 1:394-402

Cooper J (1970a) The breeding of the Fiscal Shrike in southern Africa. Ostrich 42:166-174

Cooper J (1970b) Post-nestling development in the Fiscal Shrike. Ostrich 42:175-178

Coulombe HN (1970) Physiological and physical aspects of temperature regulation in the Burrowing Owl Speotyto cunicularia. Comp Biochem Physiol 35:307-337
Coulombe HN (1971) Behavior and population ecology of the Burrowing Owl, Speotyto cunicularia, in the Imperial Valley of California. Condor 73:162-176
Cracraft J (1981) Toward a phylogenetic classification of the recent birds of the world (Class Aves). Auk 98:681-714
Craig RB (1978) An analysis of the predatory behavior of the Loggerhead Shrike. Auk 95:221-234
Craig RB, Trost CH (1979) The biology and nesting density of breeding American Kestrels and Long-eared Owls on the Big Lost River, southeastern Idaho. Wilson Bull 91:50-61
Dawson JW, Mannan RW (1991) Dominance hierarchies and helper contributions in Harris' Hawks. Auk 108:649-660
Dawson WR, Bartholomew GA (1968) Temperature regulation and water economy of desert birds. In: Brown GW Jr (ed) Desert biology, vol 1. Academic Press, New York, pp 357-394
De La Torre J (1990) Owls: their life and behavior. Crown, New York
Dittami JP, Knauer B (1986) Seasonal organization of breeding and molting in the Fiscal Shrike (Lanius collaris). J Ornithol 127:29-84
Donazar JA, Ceballos O (1990) Acquisition of food by fledgling Egyptian Vultures Neophron percnopterus by nest-switching and acceptance by foster adults. Ibis 132:603-617
Douglas MM (1990) Masters of the macabre. Birder's World 4(6):10-14
Dow DD (1980) Systems and strategies of communal breeding in Australian birds. Proc 17th Int Ornithol Congr, pp 875-881
Durango S (1956) Territory in the Red-backed Shrike Lanius collurio. Ibis 98:476-484
Edwards TC Jr, Callopy MW, Steenhof K, Kochert MN (1988) Sex ratios of fledgling Golden Eagles. Auk 105:793-796
Ehrlich PR, Dobkin DS, Wheye D (1988) Birder's Handbook. Simon and Schuster, New York
Ellis DH, Whaley WH (1979) Two winter breeding records for the Harris' Hawk. Auk 96:413
Erlinge S, Goransson G, Hogstedt G, Jansson G, Liberg O, Loman J, Nilsson I, von Schantz T, Sylven M (1984) Can vertebrate predators regulate their prey? Am Nat 123:125-133
Estrella RR (1994) Group size and flight altitude of Turkey Vultures in two habitats in Mexico. Wilson Bull 106:749-752
Evans DL (1977) Uncommon natural injuries in hawks. Auk 94:585-586
Farner DS (1967) The control of avian reproductive cycles. Proc 14th Int Ornithol Congr, pp 107-133
Farner DS, Serventy DL (1960) The timing of reproduction in birds in the arid regions of Australia. Anat Rec 137:354
Fenna L, Boag DA (1974) Adaptive significance of the caeca in Japanese Quail and Spruce Grouse (Galliformes). Can J Zool 52:1577-1584
Fisher CD, Lindgren E, Dawson WR (1972) Drinking patterns and behavior of Australian desert birds in relation to their ecology and abundance. Condor 74:111-136
Ganey JL, Balda RP, King RM (1993) Metabolic rate and evaporative water loss of Mexican Spotted and Great Horned Owls. Wilson Bull 105:645-656
Gaucher P, Thiollay JM, Eichaker X (1995) The Sooty Falcon Falco concolor on the Red Sea coast of Saudi Arabia: distribution, numbers and conservation. Ibis 137:29-34
Gaulden RL (1993) Suburban settlers. Birder's World 7(6):36-40
Gehlbach FR (1986) Odd couples of suburbia. Nat Hist 95(6):56-66
Getz LL (1961) Hunting areas of the Long-eared Owl. Wilson Bull 73:79-82
Gill FB (1990) Ornithology. Freeman, New York
Goad MS, Mannan RW (1987) Nest site selection by Elf Owls in Saguaro National Monument, Arizona. Condor 89:659-662
Goldsmith TH (1980) Hummingbirds see near ultraviolet light. Science 207:786-788
Goodman SM (1986) Identification of mammalian prey remains in owl pellets from the central Egyptian Red Sea mountains. Mammalia 50:401-403
Goodman SM, Sabry H (1984) A specimen record of the Hume's Tawny Owl Strix butleri from Egypt. Bull. Brit. Ornith. Club 104:79-84
Goodwin GA (1977) Golden Eagle predation on pronghorn antelope. Auk 94:789-790

Grimes LG (1976) The occurrence of cooperative breeding behavior in African birds. Ostrich 47:1-15

Grimes LG (1979) Sexual dimorphism in the Yellow-billed Shrike (Corvinella corvina) and in other African shrikes (Subfamily Laniidae). Bull Br Ornithol Club 99:33-36

Grimes LG (1980) Observations of group behavior and breeding biology of the Yellow-billed Shrike (Corvinella corvina). Ibis 22:166-192

Grossman ML, Hamlet J (1964) Birds of prey of the world. Bonanza Books, New York

Grubb TG (1995) Food habits of Bald Eagles breeding in the Arizona desert. Wilson Bull 107:258-274

Hamilton WJ III (1973) Life's color code. McGraw-Hill, New York

Hammond Inc (1992) Citation world atlas. Hammond, Maplewood, New Jersey

Harris T (1988) Shrikes of southern Africa. Struik Winchester, Cape Town

Harrison CJO (1971) The sunbathing of the roadrunner. Avic Mag 77(4):128

Hart JA (1971) The flocking and foraging behavior of the West African Red-bellied Shrike. Ostrich 42:294-295

Hatch DE (1970) Energy conserving and heat dissipating mechanisms of the Turkey Vulture. Auk 87:111-124

Haug EA, Didiuk AB (1993) Use of recorded calls to detect Burrowing Owls. J Field Ornithol 64:188-194

Hayes SR, Gessaman JA (1980) The combined effects of air temperature, wind, and radiation on the resting metabolism of avian raptors. J Therm Biol 5:119-125

Heath JE (1962) Temperature fluctuation in the Turkey Vulture. Condor 64:234-235

Hector DP (1985) The diet of the Aplomado Falcon (Falco femoralis) in eastern Mexico. Condor 87:336-342

Hector DP (1986) Cooperative hunting and its relationship to foraging success and prey size in an avian predator. Ethology 73:247-257

Heinrich B (1993) Kinglets' realm of cold. Nat Hist 102(2):4-9

Hils KG (1985) Foraging behavior in the Greater Roadrunner. Bachelor's Thesis, Miami University, Oxford, Ohio

Hiraldo F, Delibes M, Donazar JA (1991) Comparison of diets of Turkey Vultures in three regions of northern Mexico. J Field Ornithol 62:319-324

Hogg P, Dare PJ, Rintoul JV (1984) Palaearctic migrants in the central Sudan. Ibis 126:307-331

Houston DC (1974) Food searching in griffon vultures. East Afr Wildl J 12:63-77

Houston DC (1976) Breeding of the White-backed and Ruppell's Griffon Vultures, Gyps africanus and G. rueppellii. Ibis 118:14-40

Houston DC (1986) Scavenging efficiency of Turkey Vultures in tropical forest. Condor 88:318-323

Houston DC (1988) Digestive efficiency and hunting behavior in cats, dogs and vultures. J Zool Lond 216:603-605

Houston DC (1990) A change in the breeding season of Rüppell's Griffon Vultures Gyps rueppellii in the Serengeti in response to changes in ungulate populations. Ibis 132:36-41

Hume R (1991) Owls of the world. Running Press, Philadelphia

Hustler K, Howells WW (1988) The effect of primary production on breeding success and habitat selection in the African Hawk-Eagle. Condor 90:583-587

Hustler K, Howells WW (1989) Habitat preference, breeding success and the effect of primary production on Tawny Eagles Aquila repax in the tropics. Ibis 131:33-40

Irwin MPS (1956) Notes on the drinking habits of birds in the semidesertic Bechuanaland. Bull Br Ornithol Club 76:99-101

Jaksić FM, Braker HE (1983) Food-niche relationships and guild structure of diurnal birds of prey: competition versus opportunism. Can J Zool 61:2230-2241

Jaksić FM, Yáñez JL (1979) The diet of the Barn Owls in central Chile and its relation to the availability of prey. Auk 96:619-621

Jaksić FM, Yáñez JL, Schlatter RP (1980) Prey of the Harris' Hawk in central Chile. Auk 97:196-198

Jamieson IG (1991) The unselected hypothesis for the evolution of helping behavior: too much or too little emphasis on natural selection? Am Nat 138:271-282

Janzen DH (1977) Why fruits rot, seeds mold and meat spoils. Am Nat 111:691-713

Johnsgard PA (1990) Hawks, eagles, and falcons of North America. Smithsonian Institution Press, Washington, DC

Johnsgard PA (1991) Photo essay: Burrowing Owls. Birder's World 5(1):30-34

Jones CG, Shachak M (1994) Desert snails daily grind. Nat Hist 103(8):56-61

Kahl PM Jr (1963) Thermoregulation in the Wood Stork, with special reference to the role of the legs. Physiol Zool 36:141-151

Kaiser TJ, Bucher TL (1985) The consequences of reverse sexual size dimorphism for oxygen consumption, ventilation, and water loss in relation to ambient temperature in the Prairie Falcon, Falco mexicanus. Physiol Zool 58:748-758

Karasov WH (1990) Digestion in birds: chemical and physiological determinants and ecological implications. Stud Avian Biol 13:391-415

Karmali J (1980) Birds of Africa. Viking Press, New York

Kaufman DW (1973) Shrike prey selection: color or conspicuousness? Auk 90:204-205

Kaufman K, Stenberg K (1988) Calling all owls. Birder's World 2(2):14-18

Kavanau JL, Ramos J (1970) Roadrunners: activity of captive individuals. Science 169: 780-782

Kear J (1968) Plant poison in the diet of wild birds. Bull Br Ornithol Club 88:98-102

Keast JA, Marshall AJ (1954) The influence of drought and rainfall on reproduction in Australian desert birds. Proc Zool Soc Lond 124:493-499

Kemp AC (1991) Timing of laying by Greater Kestrels Falco rupicoloides near Pretoria, South Africa. Ostrich 62:35-39

Kendeigh SC (1961) Energy of birds conserved by roosting in cavities. Wilson Bull 73:140-147

Kendeigh SC (1970) Energy requirements for existence in relation to size of bird. Condor 72:60-65

Kendeigh SC, Blem CR (1974) Metabolic adaptation to local climate in birds. Comp Biochem Physiol 48:175-187

Kilgore DL Jr, Bernstein MH, Hudson DM (1976) Brain temperatures in birds. J Comp Physiol 110:209-215

Klem D Jr, Hillegass BS, Peters DA (1985) Raptors killing raptors. Wilson Bull 97:230-231

Knowlton GF, Harmston FC (1944) Food of White-rumped Shrikes. Auk 61:642-643

Knudsen EI (1981) The hearing of the Barn Owl. Sci Am 245(6):112-125

Komen J (1991) Energy requirements of nestling Cape Vultures. Condor 93:153-158

Konishi M (1983) Night owls are good listeners. Nat Hist 92(9):56-59

Kruuk H (1967) The competition for food between vultures in East Africa. Ardea 55:171-193

Lamberton K (1993) Air supremacy. Birder's World 7:24-27

Larochelle J, Delson J, Schmidt-Nielson K (1982) Temperature regulation in the Black Vulture. Can J Zool 60:491-494

Lasiewski RC, Bernstein MH, Ohmart RD (1971) Cutaneous water loss in the roadrunner and poor-will. Condor 73(4):470-472

Lawick-Goodall J van (1968) Tool-using birds: the Egyptian Vulture. Natl Geogr Mag 133:630-641

Laybourne RC (1974) Collision between a vulture and an aircraft at an altitude of 37,000 feet. Wilson Bull 86:461-462

Leopold AS (1967) The desert. Time-Life, New York

Ligon JD (1967) Relationships of the cathartid vultures. Occas Pap Mus Zool Univ Mich 651

Ligon JD (1968) The biology of the Elf Owl, Micrathene whitneyi. Misc Publ Mus Zool Univ Mich 136:1-70

Ligon JD (1969) Some aspects of temperature relations in small owls. Auk 86:458-472

Ligon JD, Stacey PB (1991) The origin and maintenance of helping behavior in birds. Am Nat 138:254-258

Lima SL (1993) Ecological and evolutionary perspectives on escape from predatory attack: a survey of North American birds. Wilson Bull 105:1-47

Lima SL, Valone TJ (1991) Predators and avian community organization: an experiment in a semi-desert grassland. Oecologia 86:105-112

Lustic S, Adam M, Hinko A (1980) Interaction between posture, color, and the radiative heat load in birds. Science 208:1052-1053

MacDonald MA (1980) The ecology of the Fiscal Shrike of Ghana, and a comparison with studies from southern Africa. Ostrich 51(2):65-74
Mader WJ (1977a) Harris' Hawks lay three clutches of eggs in one year. Auk 94:370-371
Mader WJ (1977b) Chemical residues in Arizona Harris' Hawk eggs. Auk 94:587-588
Mader WJ (1978) A comparative nesting study of Red-tailed Hawks and Harris' Hawks in southern Arizona. Auk 95:327-337
Mader WJ (1979) Breeding behavior of a polyandrous trio of Harris' Hawks in southern Arizona. Auk 96:776-788
Marshall BR, Cooper J (1969) Observations on the breeding biology of the Fiscal Shrike. Ostrich 40:141-149
Marti CD (1976) A review of prey selection by the Long-eared Owl. Condor 78:331-336
Marti CD, Hogue JG (1979) Selection of prey by size in Screech Owls. Auk 96:319-327
Martin AC, Zim HS, Nelson AL (1961) American wildlife and plants. Dover, New York
Mayr E (1982) The growth of biological thought. Harvard University Press Cambridge
McNutt JW (1984) A Peregrine Falcon polymorph: observations of the reproductive behavior of Falco kreyenborgi. Condor 86:378-382
McWilliams SR, Dunn JP, Raveling DG (1994) Predator-prey interactions between eagles and cackling Canada and Ross' Geese during winter in California. Wilson Bull 106:272-288
Mendelssohn H, Yom-Tov Y, Safriel U (1975) Hume's Tawny Owl Strix butleri in the Judean, Negev and Sinai Deserts. Ibis 117:110-111
Miller AH (1931) Systematic revision and natural history of the American shrikes (Lanius). Univ Calif Publ Zool 38:11-242
Miller AH, Miller L (1951) Geographic variation of the screech owls of the deserts of western North America. Condor 53:161-177
Miller GO (1988) Beep! beep!. Birder's World 2(5):16-18
Miller L (1948) An enemy of the horned lizard. Copeia 1948:67
Millsap BA, Millsap PA (1987) Burrow nesting by Common Barn-Owls in north central Colorado. Condor 89:668-670
Milne LJ, Milne MJ (1950) Notes on the behavior of horned toads. Am Midl Nat 44:720-741
Mollhagen TR, Wiley RW, Packard RL (1972) Prey remains in Golden Eagle nests: Texas and New Mexico. J Wildl Manage 36: 784-792
Moreau RE, Moreau WM (1939) Observations on some east African birds. Ibis 14:296-323
Mosher JA, White CM (1978) Falcon temperature regulation. Auk 95:80-84
Mueller HC (1967) Turkey Vultures attack living prey. Auk 84:430-431
Mueller HC (1974) Factors influencing prey selection in the American Kestrel. Auk 91:705-721
Nichol W (1964) Two males and one female of Long-tailed Shrike (Molestes melanoleucus) attending to the same nest. Ostrich 35:68-69
Nicolson SW (1990) Water relations of the Namib tenebrionid beetles. In: Seely MK (ed) Namib ecology: 25 years of Namib research. Transvaal Museum Monograph, no 7. Transvaal Museum, Pretoria, pp 173-178
Ohmart RD (1972) Physiological and ecological observations oncerning the salt-secreting nasal glands of the roadrunner. Comp Biochem Physiol 43(2):311-316
Ohmart RD, Lasiewski RC (1971) Roadrunners: energy conservation by hypothermia and absorption of sunlight. Science 172:67-69
Ohmart RD, Chapman TE, McFarland LZ (1970a) Water turnover in roadrunners under different environmental conditions. Auk 87:787-793
Ohmart RD, McFarland LZ, Morgan JP (1970b) Urographic evidence that urine enters the rectum and cecum of the roadrunner (Geococcyx californianus) Aves. Comp Biochem Physiol 35:487-490
Palmer RS (1988a) Handbook of North American birds, vol 4. Yale University Press, New Haven
Palmer RS (1988b) Handbook of North American birds vol 5. Yale University Press, New Haven
Parrish JW, Ptacek JA, Will KL (1984) The detection of near-ultraviolet light by nonmigratory and migratory birds. Auk 101:53-58
Payne RS (1971) Acoustic location of prey by Barn Owls (Tyto alba). J Exp Biol 54:535-537
Pennycuick CJ (1972) Soaring behavior and performance of some East African birds, observed from a motor-glider. Ibis 114:178-218

Petrides GA (1959) Competition between five species of East African vultures. Auk 76:104-106
Petrov MP (1975) Deserts of the world. John Wiley, New York
Pianka ER, Parker WS (1975) Ecology of horned lizards: a review with special reference to Phryosoma platyrhinos. Copeia 1975:141-162
Pilz WR, Seibert LK (1978) Fratricide and cannibalism in Swainson's Hawk. Auk 95:584-585
Pitcher EJ (1977) Nest site selection for Prairie Falcons. Auk 94:371
Poulin B, Lefebvre G, McNeil R (1993) Variations in bird abundance in tropical and semi-arid habitats. Ibis 135:432-441
Preston CR (1980) Differential perch site selection by color morphs of the Red-tailed Hawk (Buteo jamaicensis). Auk 97:782-789
Proctor NS, Lynch PJ (1993) Manual of ornithology. Yale University Press, New Haven
Reid WH, Fulbright HJ (1981) Impaled prey of the Loggerhead Shrike in the northern Chihuahuan Desert. Southwest Nat 26:204-205
Reynolds RT, Meslow EC (1984) Partitioning of food and niche characteristics of coexisting Accipiter during breeding. Auk 101:761-779
Rice WR (1982) Acoustical location of prey by the Marsh Hawk: adaptation to concealed prey. Auk 99:403-413
Ridgway R (1914) The birds of North and Middle America, part 6. Government Printing Off, Washington, DC
Russell RW (1991) Nocturnal flight by migrant "diurnal" raptors. J Field Ornithol 62:505-508
Salt GW (1964) Respiratory evaporation in birds. Biol Rev 39:113-136
Santana EC, Knight RL, Temple SA (1986) Parental care at a Red-tailed Hawk nest tended by three adults. Condor 88:109-110
Schildmacher H (1936) Vogel und Salzwasser, ein Beitrag zur Frage des Wasserhaushaltes der Vögel. Ornithol Monatsber 44:13-19
Schlatter RP, Yáñez JL, Núñez H, Jaksić FM (1980) The diet of the Burrowing Owl in central Chile and its relation to prey size. Auk 97:616-619
Schmidt-Nielsen K (1978) Desert animals: physiological problems of heat and water. Dover, New York
Schmidt-Nielsen K, Jorgensen CB, Osaki H (1958) Extrarenal salt excretion in birds. Am J, Physiol 193:101-107
Schmidt-Nielsen K, Borut A, Lee P, Crawford E Jr (1963) Nasal salt excretion and the possible function of the cloaca in water conservation. Science 142:1300-1301
Schmutz JK, Fyfe RW (1987) Migration and mortality of Alberta Ferruginous Hawks. Condor 89:169-174
Selander RK (1964) Speciation in wrens of the genus Campylorhynchus. Univ Calif Publ Zool 74: 1-305
Serventy DL (1971) Biology of desert birds. In: Farner DS, King JR, Parkes KC (eds) Avian biology, vol 1. Academic Press, New York
Sherbrooke WC (1990) Predatory behavior of captive Greater Roadrunners feeding on horned lizards. Wilson Bull 102:171-174
Sherbrooke WC, Montanucci RR (1988) Stone mimicry in the round-tailed horned lizard Phrynosoma modestum (Sauria: Iguanidae). J Arid Environ 14:275-284
Sibley CG, Monroe BL Jr (1990) Distribution and taxonomy of birds of the world. Yale University Press, New Haven
Simmons RE (1993) Effects of supplementary food on density-reduced breeding in an African eagle: adaptive restraint or ecological constraint? Ibis 135:394-402
Skutch AF (1961) Helpers among birds. Condor 63:198-226
Slack RS (1975) Effects of prey size on Loggerhead Shrike predation. Auk 92:812-814
Smith EFG, Arctander P, Fjeldså J, Amir OG (1991) A new species of shrike (Laniidae: Laniarius) from Somalia, verified by DNA sequence data from the only known individual. Ibis 133:227-235
Smith SA, Paselk RA (1986) Olfactory sensitivity of the Turkey Vulture (Cathartes aura) to three carrion-associated odorants. Auk 103:586-592
Smith SM (1972) The ontogeny of impaling behavior in the Loggerhead Shrike, Lanius ludovicianus L. Behavior 42:232-247

Snow C (1974) Ferruginous Hawk, Buteo regalis. Habitat Management Series for Unique or Endangered Species, report no 13. Bureau of Land Management, Technical Note T/N - 255

Snyder NFR, Snyder HA (1979) Biology of North American accipiters in Arizona and New Mexico. National Geographic Society Research Reports, 1970 Projects, pp 487-491

Snyder RL, Jenson W, Cheney CD (1976) Environmental familiarity and activity: aspects of prey selection for a Ferruginous Hawk. Condor 78:138-139

Soucy LJ Jr (1979) A nearly synchronous hatching of Barn Owls. Auk 96:204

Squires JR, Anderson SH, Oakleaf R (1991) Prairie Falcons quit nesting in response to spring snowstorm. J Field Ornithol 62:191-194

Stager KE (1964) The role of olfaction in food location by the Turkey Vulture (Cathartes aura). Los Ang Cy Mus Contrib Sci 81

Stark H, Liechti F (1993) Do Levant Sparrowhawks Accipiter brevipes also migrate at night? Ibis 135:233-236

Steen I, Steen JB (1965) The importance of the legs in the thermoregulation of birds. Acta Physiol Scand 63:285-291

Stensrude C (1965) Observations on a pair of Gray Hawks in southern Arizona. Condor 67:319-321

Storer RW (1987) Avian exotica: the Secretary-bird. Birder's World 1(6):52-53

Storer RW (1988) Avian exotica: the Egyptian Vulture. Birder's World 2(6):58-59

Sutton GM (1929) Tactics of the domestic pigeon in evading the Duck Hawk. Wilson Bull 41:41-42

Tanner WW, Krogh JE (1973) Ecology of Phrynosoma platyrhinos at the Nevada test site, Nye County, Nevada. Herpetologica 29:327-342

Tarboton W (1963) Breeding observations on the Red-billed Helmet Shrike. Bokmankierie 15:1-3

Taylor C, Lillis C, LeMone P (1989) Fundamentals of nursing. JB Lippincott, Philadelphia

Taylor RL, Temple SA, Bird DM (1991) Nutritional and energetic implications for raptors consuming starving prey. Auk 108:716-719

Terres JK (1980) The Audubon Society Encyclopedia of North American Birds. Knopf, New York

Thiollay JM, Clobert J (1990) Comparative foraging adaptations of small raptors in a dense African savanna. Ibis 132:42-57

Thomas DH (1982) Salt and water excretion by birds: the lower intestine as an integrator of renal and intestinal excretion. Comp Biochem Physiol 71(4):527-536

Thurow TL, White CM (1983) Nest site relationship between the Ferruginous Hawk and Swainson's Hawk. J Field Ornithol 54:401-406

Toops C (1994) On pins and needles. Birder's World 8(1):17-21

Torre-Bueno JR (1978) Evaporative cooling and water balance during flight in birds. J Exp Biol 75:231-236

Tortora GJ, Anagnostakos NP (1984) Principles of anatomy and physiology, 4th edn. Harper and Row, New York

Tucker VA (1991) Stereoscopic views of three-dimensional, rectangular flight paths in descending African White-backed Vultures (Gyps africanus). Auk 108:1-7

Turner FB, Medica PA (1982) The distribution and abundance of the flat-tailed horned lizard (Phrynosoma mcallii). Copeia 1982:815-823

Varland DE, Klaas EE, Loughin TM (1993) Use of habitat and perches, causes of mortality and time until dispersal in post-fledging American Kestrels. J Field Ornithol 64:169-178

Vehrencamp SL (1982) Body temperatures of incubating versus non-incubating roadrunners. Condor 84(2):203-207

Voet D, Voet JG (1990) Biochemistry. John Wiley, New York

Wagner FH (1980) Wildlife of the deserts. Abrams, New York

Wakeley JS (1978) Hunting methods and factors affecting their use by Ferruginous Hawks. Condor 80: 327-333

Walker EP (1975) Mammals of the world. Johns Hopkins University Press, Baltimore

Walsberg GE (1982) Coat color, solar heat gain, and conspicuousness in the Phainopepla. Auk 99:495-502

Walsberg GE (1986) Thermal consequences of roost-site selection: the relative importance of three modes of heat conservation. Auk 103:1-7
Walsberg GE (1990) Communal roosting in a very small bird: consequences for the thermal and respiratory gas environments. Condor 92:795-798
Walsberg GE, Campbell GS, King JR (1978) Animal coat color and radiative heat gain: a re-evaluation. J Comp Physiol 126:211-222
Wasser JS (1986) The relationship of energetics of falconiform birds to body mass and climate. Condor 88:57-62
Weathers WW, Buttemer WA, Hayworth AM, Nagy KA (1984) An evaluation of time-budget estimates of daily energy expenditure in birds. Auk 101:459-472
Weigand JP (1967) Ferruginous Hawk attacks Great Horned Owl. Auk 84:433
Welty JC (1982) The life of birds. Saunders, Philadelphia
Whitman FN (1924) Nesting habits of the Long-eared Owl. Auk 41:479-480
Williams JB (1996) A phylogenetic perspective of evaporative water loss in birds. Auk 113:457-473
Williams JB, Plessis MA du, Siegfried WR (1991) Green Woodhoopoes (Phoeniculus purpureus) and obligate cavity roosting provide a test of the thermoregulatory insufficiency hypothesis. Auk 108:285-293
Wilson RT, Wilson MP (1983) Notes on a Gray Kestrel Falco ardosiaceus brood in central Mali. Bull Br Ornithol Club 103:118-119
Windingstad RM, Stiles HE, Drewien RC (1981) Whooping Crane preyed upon by Golden Eagle. Auk 98:393-394
Winterbottom JM (1971) The birds of the Deka expedition. Ostrich Suppl 9:5-26
Withers PC, Williams JB (1990) Metabolic and respiratory physiology of an arid-adapted Australian bird, the Spinifex Pigeon. Condor 92:961-969
Woffinden ND, Murphy JR (1983) Ferruginous Hawk nest site selection. J Wildl Manage 47:216-219
Woods CP (1993a) Variation in Loggerhead Shrike nest composition between two shrub species in southwest Idaho. J Field Ornithol 64:352-357
Woods CP (1993b) Parent Loggerhead Shrikes induce nestlings to fledge. Wilson Bull 105:531-532
Yosef R (1993) Prey transport by Loggerhead Shrikes. Condor 95:231-233
Yosef R, Grubb TC Jr (1993) Effect of vegetation height on hunting behavior and diet of Loggerhead Shrikes. Condor 95:127-131
Yosef R, Grubb TC Jr (1994) Resource dependence and territory size in Loggerhead Shrikes (Lanius ludovicianus). Auk 111:465-469
Yosef R, Pinshaw B (1988a) Nesting transfer in the Northern Shrike (Lanius excubitor). Auk 105:580-581
Yosef R, Pinshaw B (1988b) Polygyny in the Northern Shrike (Lanius excubitor) in Israel. Auk 105:581-582
Yosef R, Pinshaw (1989) Cache size in shrikes influences female mate choice and reproductive success. Auk 106:418-421
Zack S (1986) Behavior and breeding biology of the cooperatively breeding Grey-backed Fiscal Shrike Lanius excubitorius in Kenya. Ibis 128:214-233
Zack S, Ligon JD (1985) Cooperative breeding in Lanius shrikes, I. Habitat and demography of two sympatric species. Auk 102:754-765
Zimmerman DA (1970) Roadrunner predation on passerine birds. Condor 72(4):475-476

Species Index

Subject Index

Springer and the environment

At Springer we firmly believe that an international science publisher has a special obligation to the environment, and our corporate policies consistently reflect this conviction.
We also expect our business partners – paper mills, printers, packaging manufacturers, etc. – to commit themselves to using materials and production processes that do not harm the environment. The paper in this book is made from low- or no-chlorine pulp and is acid free, in conformance with international standards for paper permanency.

Springer

Druck: Strauss Offsetdruck, Mörlenbach
Verarbeitung: Schäffer, Grünstadt